Vocabulary to Teach

The content of the stories in *Read and Understand, Science, Grades 4–6* requires that specific science vocabulary be used. These words and additional words that your students may not know are given below. It is advisable to read the story in advance to pinpoint any other vocabulary that should be introduced.

The Common Cold
microscope, virus, alights, cell, infected, receptors, blood vessels, chemicals, antibodies, miserable, agreeable, specifically, mucous

Everybody into the Pool!
horizon, intertidal zone, hermit crab, exoskeleton, abdomens, tide pools, echinoderm, rays, suction cups, hinged, clam, mussel, scallop, bivalves, siphon, algae, dulse, air bladders, rhythm, holdfasts, barnacles, lobster, sand dollar, sea urchin, glistening, cellophane, scrambling, protection, overpopulate, lurking, environment

Why Do Basketballs Bounce? Why Do Eggs...Not?
energy, transfer, potential energy, kinetic energy, molecules, powerful

The Sun's Radiation
radiant energy, radiates, ultraviolet waves, ultraviolet energy, photosynthesis, atmosphere, ultraviolet radiation

Gravity
gravity, solar system, orbiting, mass, matter, gravitational pull, dense, ebb and flow, bulge, construction paper, exert

Bouncing and Bending Beams of Light
absorbs, reflected, reflection, angle, mirror image, regular reflection, reversed, refraction, obvious, experimenting

A Better Gizmo
cylinders, vibrations, amplified, sketch, breadboard, model, prototype, gizmos, complicated, motto, auditorium, contraption, practical, calculators, audience, improved, infinite, phonograph, Thomas Edison, Emile Berliner, gramophone, necessarily, consider, balsa wood

All Rise
organism, fungus, fungi, budding, chlorophyll, substances, enzymes, proteins, alcohol, carbon dioxide, ferments, fermentation, gluten, semi-dormant, nutritional yeast, antibiotics, B-complex vitamins, fortified, commercial, Louis Pasteur

Making Your Car Move
boiler, pressure, petroleum, battery, recharged, renewable energy sources, natural resources, pollution, smog-free, solar-powered, solar panels, hybrid electric vehicles, biomass, gasohol, popular, alternative, environmental, advantages, efficient, convert, assure, transferred

Comets
nucleus, vapor, coma, oval-shaped orbits, Halley's comet, gravity, comet Shoemaker-Levy, comet Hale-Bopp, research, reflect, kilometers, occurred

Asking Mr. Aims
rock quarry, limestone, steel, cement, iron ore, carbon, lime, dynamite, soil additives, slate, quartz, coal, granite, erupts, magma, lava, obsidian, pumice, igneous, sedimentary, metamorphic, weathered, layered, pressure, quartzite, backhoe, factory, antacids, marble, sculptures

The Greatest Show on Earth
veins, arteries, orbits, astonishing, total eclipse, fiery atmosphere, corona, partial phases, totality, fabulous, hemisphere, waltz, spectacular, volunteer, star-spangled, position

A Compound Mystery
chemical compounds, mixture, substances, properties, characteristics, liquid, sodium, ammonia, chlorine, Osgood Blastwood, Inspector Yoghurty, Claudia Simpson, cryptically, suicide, stupidity, explosive combination, original, resulting, poisonous, mackerel, apparently, medical examiner

A Breath of Wind
cumulonimbus, upper-level air, air masses, front, thunderheads, tornado alley, supercell, mesocyclones, Oklahoma, Gulf of Mexico, Canada, Rockies, colliding, eerie, reformed, dissolved, funnel shape, inhaling, counterclockwise, horrified, fascinated, demolish, geraniums, wrenching, bellowed, hickory, squall

An Inside Look at the San Andreas Fault
San Andreas Fault, Earth's crust, tectonic plates, molten mantle, trenches, Pacific Plate, North American Plate, vegetation, terrain, observers, geologist, collide, San Francisco, San Bernardino, continuous, lurch, vibrations, Andrew Lawson, occurred, San Diego, disaster, scenery

Dear Professor Parsec
escape velocity, gravity, speedometer, astronomer, inertia, thrust, liquid oxygen, liquid hydrogen, chemicals, launch pad, exhaust, star bound, Star Commander Lance Redshift, attractive, tempted, Sir Isaac Newton, Laws of Motion, suped-up, combination, enormous, stages, cruise

Jocelyn Bell Burnell
radio telescope, radio signals, antennae, scan, scruff, static, pulses, aliens, pulsar, neutron star, Nobel Prize, astronomy, gamma-ray and X-ray telescopes, satellites, Jocelyn Bell Burnell, challenge, Irish-born, Cambridge University, England, professor, collapses, dense, career, interference

Probing the Periodic Table
elements, hydrogen, oxygen, silicon, atoms, atomic mass, regular intervals, periodic table, atomic number, protons, electrons, Dmitry Mendeleev, neutrons

The Mystery of the Melted Candy Bar
engineer, patented, magnetron, microwaves, radar, Radarange, molecules, National Inventors Hall of Fame, Percy Spencer, orphaned, devised, Nazi, mass-produce

Just for the Fun of It
aerodynamics, leading-edge flap, state-of-the-art, U.S. Patent Office, Lacey's Cross-Road, Al Glover, Depression, rural Alabama, model airplanes, balsa wood, glide, coaxing, daring stunts, maneuvering, nose dives, tinkered, Chrysler

Global Warming
experts, atmosphere, Celsius, global warming, greenhouse gases, natural cycle, carbon dioxide, fossil fuels, solar energy, solar technology, dramatic effects, climate, affect, solution, commit

Here's Looking at You, Kid
microscopic, eyespots, planarian, scallop, predators, predatory, prey, hermit crab, stalks, nocturnal, vision, primitive, sensitive, complex, improvement, developed, ability, detect, arranged, unique, motion, position, scrounges

Understanding Extinction
extinctions, marine species, fossils, competition, evolution, climate, natural disasters, theory, natural resources, endangers, survival, animal species, ratio, predators, prey, legal hunting concessions, advertisement, scuba, tropical, stalking, culture, marvel, professional, communities, rely on, trophy, tourism, exploiting, continent, appalling, safari, poacher, intruding, vacationers, extinction

McDonald Observatory
observatory, stargazing, astronomers, light pollution, Otto Struve telescope, reflector telescope, Harlan-Smith telescope, NASA, atmospheres, Viking mission, Voyager, solar system, Hobby-Eberly telescope, Lunar Ranging Station, laser beams, reflectors, satellites, Galileo Galilei, refractor, guidebook, analyze, devices

Cells: Structure and Function
bacteria, organisms, microscope, microscopic, structure, function, environment, nerve, extend, cell membrane, pores, cell wall, nucleus, genetic material, generation, nuclear membrane, cytoplasm, organelles, ribosomes, mitochondria, vacuoles, endoplasmic reticulum, proteins, convert, producing, particular, responsible, gel, designed, trillion, barrier, specially, cells

Crooked Cells
diagnosis, capillaries, hemoglobin, molecule, sickle-cell anemia, symptoms, mutation, gene, inherits, desperately, tenderness, wracked, consulted, dismay, episodes, suffered, intense pain, severe, identified, medical conference, faulty, fibers, crescent, error

Communicating Through the Ages
information exchange, communication, Internet, graphite, pulses of electric current, telegraph, typewriter, word processor, patents, transmitted, phonograph, cylinder, etches, records, antenna, broadcast, motion picture, transmits, transmitter, satellite, microchip, cellular phone, transmit radio signals, compact disc (or CD), digital recording, Johann Gutenberg, Samuel Morse, Alexander Graham Bell, Latham Sholes, Thomas Edison, Guglielmo Marconi, Valdemar Poulsen, John Baird, Tim Berners-Lee, Stephen Wozniak, access, options, misinterprets, monks, hoax, device, commercial, evolved, version, Defense Department, artificial, Tokyo, erasable

The Common Cold

It's small. In fact, it's so small that you can't see it without a microscope. Yet, it is so powerful that it can make a grown person miserable in less than a day.

Is it alive? Hmm, that's a good question. Scientists struggle with this issue, but there is no clear answer. Certainly, it isn't alive in the sense that you are alive. It isn't made up of cells. And it can multiply only when inside a cell, such as the cells in your nose and throat.

Have you guessed what it is? Yes, that's right—it's a **virus**. Specifically, it's a common cold virus.

This cold virus is an agreeable sort. It goes where people send it. If Maria carries it on her hand, then touches your hand, it stays on your hand. It stays, that is, until you scratch your nose. Then, it alights there, just inside your nostril. Uh-oh. You sniff, moving it into your nose. Oh, my. Now you are in trouble.

The virus will attach itself to a **cell** at the back of your nose or throat. Although it is much smaller than the cell, it will destroy the cell quickly. It will enter the cell, and once inside, it will make many more virus particles just like itself. The cell will become **infected** and die. When it does, it will release virus particles into your nose, where they will infect many more cells. Then they, in turn, will infect others. Before the day is over, your throat will feel rough, and your head will ache.

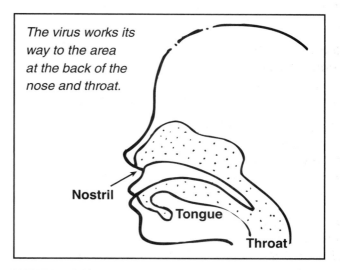

The virus works its way to the area at the back of the nose and throat.

Nostril

Tongue

Throat

Fight back, you want to tell the cells. It is a tiny virus, much tinier than you. Don't let it beat you! Your body does fight back, by making **blood vessels** bigger. The bigger vessels release special chemicals that fight the virus. Your body also starts making extra **mucous** to help wash away the virus. The swollen blood vessels and extra mucous cause your throat and your head to hurt. When your nose senses all that stuff in there that doesn't belong, you sneeze. Then you sneeze some more.

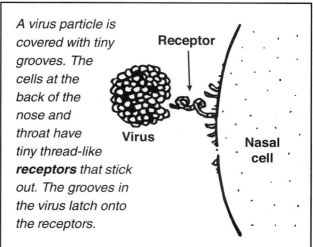

A virus particle is covered with tiny grooves. The cells at the back of the nose and throat have tiny thread-like receptors that stick out. The grooves in the virus latch onto the receptors.

Receptor

Virus

Nasal cell

Some of the mucous may wash into your lungs, carrying the virus with it. You cough to get rid of it. Your bones ache too. That's your body's way of telling you that something is wrong.

So your body is fighting back. But who is it fighting—you or the virus? Everything your body does to fight the virus seems to make you feel worse. It isn't really the tiny virus that causes you to feel bad when you have a cold. It's the ways your body reacts to it.

Why doesn't your body do a better job of fighting off the cold virus? No one really knows. This is just one of many things that scientists don't understand about colds.

Your body makes antibodies, so you can't catch the same virus twice.

One thing they do know is this: You can't catch the same cold twice. There are lots of different kinds of cold viruses. Each one is just a little bit different. Every time you catch one, your body makes antibodies that fight the virus. If the same virus comes back, your body knows how to make the **antibodies** that fight that virus very quickly, so the virus doesn't make you sick a second time. This is why young children catch colds more often than adults. Their bodies don't recognize as many different kinds of viruses.

To avoid getting cold viruses, wash your hands with soap often. Avoid sharing food or drinks with people who have colds. To avoid spreading colds, cover your mouth when you cough, and use a tissue when you sneeze. The best way to fight colds is to avoid getting them in the first place.

Questions about
The Common Cold

Answer the following questions in complete sentences.

1. Describe how the cold virus gets into your body.

_____

_____

_____

_____

2. What happens to a cell after a virus attaches to it?

_____

_____

3. What is the purpose of the extra mucous that your body makes when you have a cold?

_____

_____

4. Why do you cough and sneeze when you have a cold?

_____

_____

5. Why can't you catch the same cold virus twice?

_____

_____

_____

6. Describe two things you can do to avoid getting colds.

_____

_____

_____

Vocabulary

Synonyms are words that mean the same or almost the same thing.

A. Choose a synonym for each of the words below. Write the synonyms on the lines.

Synonyms					
responds	unhappy	settles on	stick to	easy-going	kill

1. miserable _____

2. agreeable _____

3. alights _____

4. destroy _____

5. reacts _____

6. attach _____

B. Write words from the Word Box next to the definitions below. Use clues from the article to help you.

Word Box					
particle	antibody	infected	receptor	cell	mucous

1. the smallest unit of life; the smallest living part of the body _____

2. the fluid made to wash foreign objects out of the body _____

3. damaged or made sick by a virus or bacteria _____

4. something that receives or catches something else _____

5. a substance made in the body to act against a virus _____

6. a speck, or very small bit _____

Write a Story

Write a story about a person who is catching a cold. Your story may be true, or you may make it up. Include details about how the person catches the cold, and describe how the person feels. Be sure your story has a beginning, a middle, and an end.

Everybody into the Pool!

It is early morning. The sun sits low on the horizon, still half underwater, it seems. The tide has just gone out, and the rocks are wet and glistening. The gulls cry, and their voices echo in the cool, quiet air.

Carefully, we make our way over the slippery rocks along the shore. Seaweed clings like sticky bits of green cellophane.

"There's nothing here," you say. "It's boring."

"Are you kidding?" I cry. "This is the **intertidal zone**, where the sea meets the land. There are millions of animals."

All at once, we catch sight of a brightly colored shell scrambling toward us on skinny legs.

"Wow!" you cry. "What's that?"

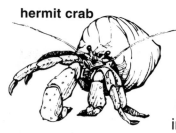
hermit crab

"It's a **hermit crab**. Hermit crabs are probably the wackiest-looking residents of the intertidal zone. They have large claws and a hard covering called an **exoskeleton** over the upper parts of their bodies, but their soft **abdomens** have no protection at all. So hermit crabs rent the shells that other animals leave behind. When they find an empty shell that seems to be just the right size, they move in. But as the crab grows, its shell house slowly becomes tighter and tighter."

"Does it get stuck in there?" you ask.

"No, the hermit crab just moves out and finds a bigger shell. Sometimes there are plants growing on the shell. Then the hermit crab looks as though it is wearing a big, fancy hat."

We walk a little farther. When the tide goes out, some of the water is left behind. It is trapped in between the rocks, forming shallow little pools. At first, you may not think there's much going on in these **tide pools**, but they're as crowded as cities. We stop and kneel down to take a closeup look.

sea star

"Look! A starfish!"

"That's what most people call it, but it's not really a fish. The correct name for it is sea star. This amazing animal is an **echinoderm**. It has five arms, called **rays**, and its mouth is underneath its body. Go ahead. Pick it up."

You reach into the cold, clear water and gently take hold of the sea star. "Hey! It's stuck."

"That's because the sea star's arms are lined with hundreds of tiny suction cups. Sea stars love to eat the clams and mussels that also live in the tide pool, but very tough shells protect these animals. The shells are hinged, like a door. A powerful muscle holds the shell closed, making it difficult for the sea star to get at the meat. The sea star has to pull really hard to open the shell. That's where the suction cups come in. The sea star climbs

10

onto the clam and wraps its arms around the top and bottom of the shell. The suction cups help the sea star pry the shell open."

"I feel a little bad for the poor clam," you say.

"But it has to be that way. If something didn't eat the clams and mussels, they would **overpopulate** the tide pool, and before long, there would be no room for any of the other animals. We see a few clams resting on the bottom of the tide pool. They don't seem to know the sea star is lurking nearby."

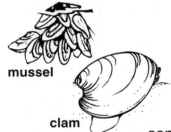

mussel

clam

"Do the clams eat anything?" you want to know. "I don't see a mouth."

"Everything eats something, but not every animal needs a mouth. Clams and mussels are **bivalves**. Bivalves filter tiny bits of food from the water. They open their shells just a little bit, and a tube called a **siphon** sucks in the water and food.

"Many colored **algae** also live in the intertidal zone. They have a strange beauty, like the plants in a science fiction movie.

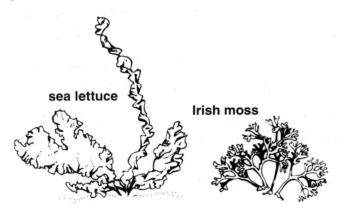

sea lettuce

Irish moss

"Look. This is sea lettuce, and this is Irish moss. And this bright purple alga is named dulse. You can eat dulse, just like a vegetable. It's pretty good."

"This is weird," you say. "What is it?"

"That's rockweed. Rockweed has tiny balloons called **air bladders** all along its stems. The stems float because of the air in the bladders."

rockweed

We stand up and stretch our legs. We can hear the waves lapping softly along the shoreline. Slowly, the sea is returning. The tide is coming in.

"The creatures that live in the intertidal zone must know how to adapt to their changing environment. Twice each day they are flooded, and twice each day many are left high and dry. It's the rhythm of the ocean, and everything that lives here must know how to dance to the rhythm. They must know how to hang on when the waves crash over them. The **barnacles** use a special cement. The seaweeds use something called **holdfasts**."

"That's a good name," you say.

barnacles

"So what did you think of the tide pools?" I ask as we make our way back across the beach.

"Pretty good," you say. "Do you think we can come again?"

Overhead, a gull screeches.

"Did you hear that?" I whisper. "I think he said 'Yes.'"

Questions about
Everybody into the Pool!

Answer the following questions in complete sentences.

Pretty Clammy

1. Why are clams so hard to open?

_____

_____

2. How do clams eat?

_____

_____

3. Sea stars love to eat clams. How are they able to get them open?

_____

_____

Pretty Sticky

4. What keeps barnacles stuck to the rocks when the waves break over them?

_____

_____

Pretty Funny

5. Why does the hermit crab move in and out of different shells during its life?

_____

_____

Just Plain Pretty

6. How do tide pools form?

_____

_____

_____

Vocabulary

Are you a good marine biologist? Can you identify these tide pool plants and animals and their characteristics? You can return to the tide pool walk to help with your identification.

1. Which of these animals is an echinoderm?
   - ◯ clam
   - ◯ hermit crab
   - ◯ sea star
   - ◯ barnacle

2. Which of these has air bladders?
   - ◯ sea lettuce
   - ◯ rockweed
   - ◯ dulse
   - ◯ Irish moss

3. Which of these is the bivalve?
   - ◯ clam
   - ◯ sand dollar
   - ◯ sea star
   - ◯ lobster

4. Which of these is the alga?
   - ◯ barnacle
   - ◯ scallop
   - ◯ sea lettuce
   - ◯ mussel

5. Which of these animals has an exoskeleton?
   - ◯ clam
   - ◯ barnacle
   - ◯ mussel
   - ◯ hermit crab

Everybody into the Pool!

This picture shows the plants and animals you "saw" in the tide pools and along the shore when you took your walk. Their names are in the box next to a number. Number the plants and animals you find. (You can look back at the story if you need some help.)

1. barnacle	3. hermit crab	5. rockweed	7. sea star
2. clam	4. mussel	6. sea lettuce	

Why Do Basketballs Bounce?
Why Do Eggs...Not?

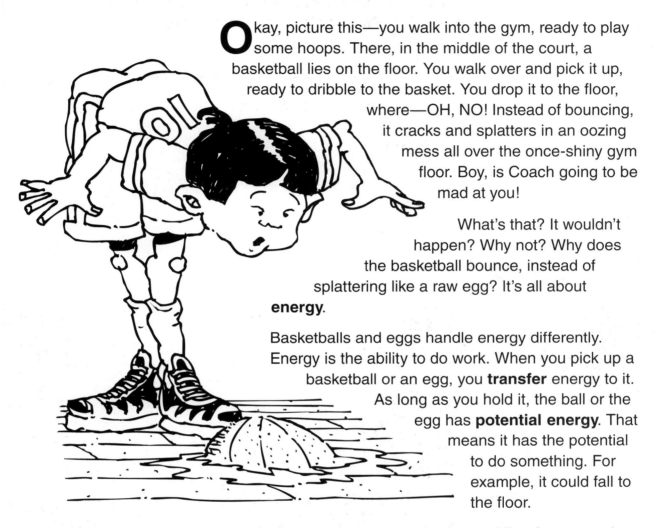

Okay, picture this—you walk into the gym, ready to play some hoops. There, in the middle of the court, a basketball lies on the floor. You walk over and pick it up, ready to dribble to the basket. You drop it to the floor, where—OH, NO! Instead of bouncing, it cracks and splatters in an oozing mess all over the once-shiny gym floor. Boy, is Coach going to be mad at you!

What's that? It wouldn't happen? Why not? Why does the basketball bounce, instead of splattering like a raw egg? It's all about **energy**.

Basketballs and eggs handle energy differently. Energy is the ability to do work. When you pick up a basketball or an egg, you **transfer** energy to it. As long as you hold it, the ball or the egg has **potential energy**. That means it has the potential to do something. For example, it could fall to the floor.

When you drop it, that potential energy changes into kinetic energy. **Kinetic energy** is the energy of motion.

When the basketball hits the floor (here's the part that's different from the egg), it stores energy as the ball dents slightly. For that brief moment, the ball has potential energy again. The **molecules** in the rubber stretch and bend, storing energy. Then the molecules go back to their original shape. The potential energy becomes kinetic energy again. The ball bounces upward.

The egg, of course, is different. Eggshell is not made of stretchy, bendy molecules that store energy. When the egg hits the floor, it smashes and breaks.

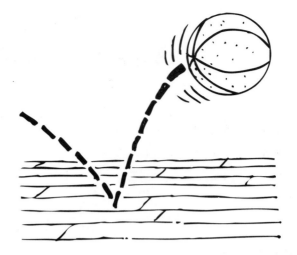

Now let's talk about the floor. What do you think happens to the floor when the basketball hits it? I'll give you a hint; it's a lot like what happens to the ball. The floor dents very slightly, storing energy for a brief moment. Then it goes back to its original shape. Most of the energy returns to the ball.

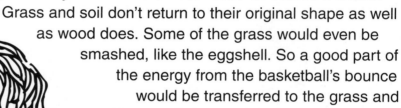

But what do you think would happen if the gym floor was covered in grass and soil, like a football field? Would it change the way the basketball bounces? Of course it would! Grass and soil don't return to their original shape as well as wood does. Some of the grass would even be smashed, like the eggshell. So a good part of the energy from the basketball's bounce would be transferred to the grass and soil. As a result, the ball wouldn't bounce as high as it would on a gym floor.

Okay, now picture this—you walk onto the tennis court, your racket in hand. You throw the ball into the air and swing your racket up for a powerful serve, when—OH, NO! Instead of shooting across the net, the ball splatters your racket, your arm, and your hair with yucky tennis ball guts. What's that? It wouldn't happen? Why not? Energy.

Questions about
Why Do Basketballs Bounce?
Why Do Eggs...Not?

Answer the following questions in complete sentences.

1. What is potential energy?

   _____

   _____

2. What is kinetic energy?

   _____

   _____

3. Briefly explain what happens when the basketball hits the floor.

   a. What happens to the basketball?

   _____

   _____

   _____

   b. What happens to the energy that caused it to fall?

   _____

   _____

   c. What happens to the floor?

   _____

   _____

4. How is what happens when the egg hits the floor different from what happens when the basketball hits the floor?

   _____

   _____

   _____

Name _____

Vocabulary

Use the action words from the article to complete the sentences. Then write new sentences using the action words.

Action Words					
dents	splatters	oozing	bounces	cracked	smashed

1. The roof of the car _____ when the hail strikes it.

   New sentence:_____

   _____

2. Armand stepped on his glasses and _____ them.

   New sentence:_____

   _____

3. The way my brother _____ around, you'd think he was made out of rubber.

   New sentence:_____

   _____

4. Green foam is _____ out of the test tube.

   New sentence:_____

   _____

5. See how the rain _____ on the windshield?

   New sentence:_____

   _____

6. As the ice _____ under her weight, she heard a loud pop.

   New sentence:_____

   _____

Name _____

Observation and Conclusion

See for yourself how a ball bounces on different surfaces.

Materials
- a rubber ball
- a measuring tape or yardstick

Procedure
1. Working with a partner, measure how high the ball bounces on three different surfaces, such as carpet, hard floor, and grass.

2. Drop the ball from shoulder height, as your partner watches the bounce and holds out a hand to mark the highest point. Use a measuring tape or yardstick to measure the height of the bounce. Record your results and then answer the questions below.

	Type of Surface	Height of Bounce
Surface 1		
Surface 2		
Surface 3		

1. On which surface did the ball bounce the highest? Why?

_____

_____

_____

2. On which surface did the ball bounce the lowest? Why?

_____

_____

_____

The Sun's Radiation

Study the pictures on this page and the following page. Read the captions. Then chant the poem, clapping your hands every time you say, "Mm-hm."

Radiant energy—that's the kind
The Sun gives off all the time.
People on Earth can only see
A part of our Sun's energy.
We see light.
Mm-hm, mm-hm, that's right.
Mm-hm, mm-hm.

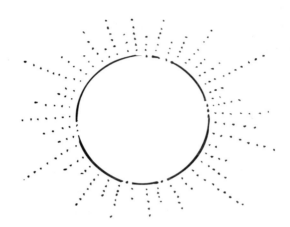

Radiant energy *is energy that radiates, or travels out in all directions. The Sun's radiant energy travels in waves, like radio waves. The waves come in many different lengths. This is what makes* *ultraviolet waves* *different from light.* *Ultraviolet energy,* *often called UV energy, travels in very short waves, shorter than light.*

Sunlight shines on Earth and you,
Light that makes the sky look blue,
Light that helps our gardens grow
With photosynthesis, you know.
That's light.
Mm-hm, mm-hm, that's right.
Mm-hm, mm-hm.

Plants use light from the Sun to help them turn chemicals from the air and ground into food for the plant. This process is called *photosynthesis.*

Sunlight helps all plants around
So hungry creatures can chow down.
Then we eat the creatures and plants,
So we'll have strength to run and dance.
Here's our chance!
Mm-hm, mm-hm, let's dance!
Mm-hm, mm-hm.

Read and Understand, Science • Grades 4–6 • EMC 3305

Another kind of wave that we don't see
Is ultraviolet energy.
It's mostly blocked before it gets here
By our Earth's upper atmosphere.
That's good,
Mm-hm, hm-hm, for the 'hood.
Mm-hm, mm-hm.

Those UV waves, so nasty and strong,
Would burn us up before too long.
The small amount that does get through
Makes lots of trouble for me and you.
Sun's up?
Mm-hm, mm-hm, cover up!
Mm-hm, mm-hm.

UV and light are only two
Radiation types, it's true.
The Sun gives off other types, I know,
But time is up, I've got to go.
I'm gone,
Mm-hm, mm-hm, so long!

*Earth's **atmosphere** blocks most of the Sun's **ultraviolet radiation**. Some of it is also lost out in space. The small amount of ultraviolet radiation that does get through to Earth's surface is believed to cause sunburn, wrinkles, and even cancer.*

Questions about
The Sun's Radiation

Answer the following questions in complete sentences.

1. What is the poem *The Sun's Radiation* mainly about?

_____

_____

2. What is radiant energy?

_____

_____

3. What part of the Sun's energy can people see?

_____

4. What is photosynthesis?

_____

_____

5. What is ultraviolet energy?

_____

_____

6. Explain why only a small amount of ultraviolet energy reaches Earth's surface.

_____

_____

_____

7. Name three negative effects of ultraviolet energy, as described in *The Sun's Radiation*.

_____

_____

_____

Vocabulary

Words that sound the same but have different meanings are called **homophones**.
Find a homophone in the poem for each of the following words. Write the homophone
and use it in a sentence.

1. son: The young mother took her five-year-old son to the zoo.

   Homophone: _____

   Sentence: _____

2. wheel: Tomás took the bent wheel off his bike.

   Homophone: _____

   Sentence: _____

3. sea: Within an hour, the ship was out on the open sea.

   Homophone: _____

   Sentence: _____

4. too: Please pass me the broccoli. Oh, pass the carrots too.

   Homophone: _____

   Sentence: _____

5. write: Would you write your name at the top of the page?

   Homophone: _____

   Sentence: _____

6. hear: The bell rang, but they couldn't hear it.

   Homophone: _____

   Sentence: _____

Name _____

Using a Venn Diagram

A Venn diagram can be helpful in organizing information, especially when you want to think about how things are alike and how they are different. Use information from the poem and captions to complete the Venn diagram below.

Write facts that are only true about light on the left side. Write facts that are only true about ultraviolet energy on the right side. Write facts that are true about both in the middle section.

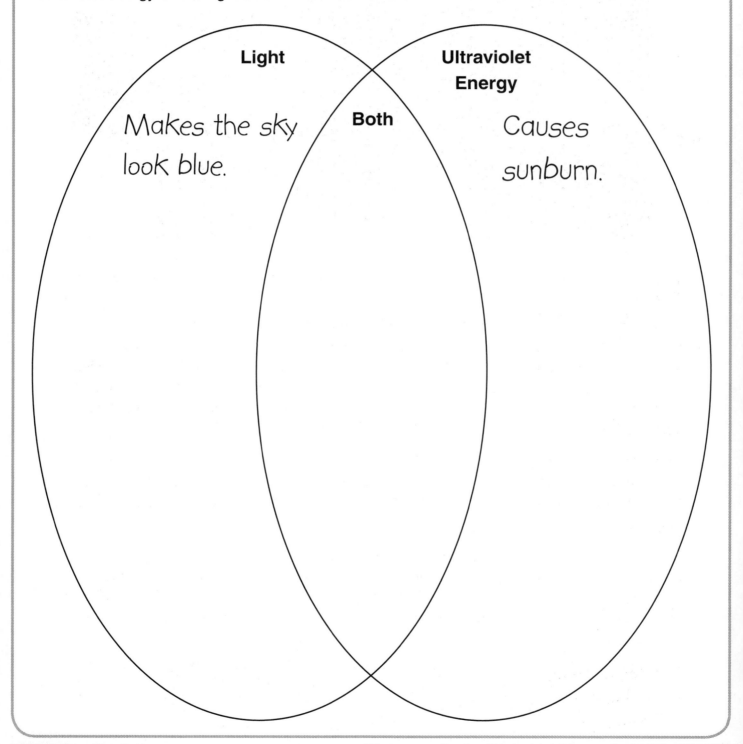

Light

Makes the sky look blue.

Both

Ultraviolet Energy

Causes sunburn.

At the end of this article, you will be asked to make a poster to illustrate what you have learned about gravity. You will need these materials:

- posterboard
- markers
- newspapers, magazines

- glue or glue sticks
- construction paper in a variety of colors
- small stick-on labels

Gravity

Every object pulls on every other object. That pulling force is known as **gravity**. The Sun and every planet in our **solar system** exert pulling forces on each other. This is what keeps the planets **orbiting** around the Sun.

Even the smallest objects exert pulling forces on each other. Gravity exists between you and your chair. It even exists between your pencil and your books! The more mass two objects have, the greater the force of gravity between them. **Mass** is the amount of **matter** in an object. The Earth contains a lot of matter. So, the force of gravity between you and Earth is great.

The distance between objects also affects the force of gravity that objects exert on each other. As objects move farther apart, the force of gravity between them lessens. If this weren't true, you might be pulled off Earth by the **gravitational pull** of the much larger Sun.

Gravity makes things fall when you drop them. It also gives them weight. It holds you on Earth's surface, and it holds the Moon in orbit around Earth. It makes a basketball swoosh down through a hoop, and a baseball drop into the glove of an outfielder. Gravity creates waterfalls and makes rivers flow down to the ocean. It makes rain and snow fall from clouds to the ground.

Gravity acts in ways that are hard to see as well. It causes dense cold air to sink—pushing hot air upward. This is why houses are often warmer upstairs than downstairs. Gravity also causes ocean tides to ebb and flow.

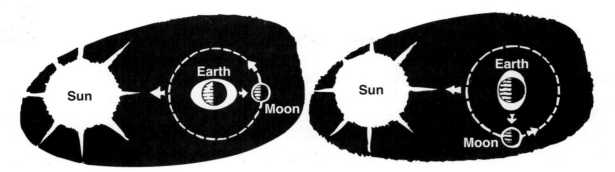

Water in the oceans moves in response to the force of gravity between Earth and the Moon. This creates a bulge in the ocean where Earth is nearest the Moon. It creates a bulge on the opposite side of Earth as well. The bulges move as Earth rotates, staying on the side of Earth that is closest to the Moon, and on the side opposite the Moon. These bulges are seen on Earth as high tides. Gravity between Earth and the Sun creates bulges too. However, the Sun is much farther away from Earth than the Moon is. So the force of gravity between Earth and the Sun creates a smaller bulge, or tide.

Gravity is at work everywhere. Look around you. Do you see examples of gravity in your classroom or home? Can you think of other examples of how gravity affects the world that you live in? Jot down a list of your ideas on the lines below.

_____

_____

_____

_____

_____

_____

_____

Share your ideas with others. Listen to what they have to say about your ideas and about ideas of their own. Then, on another sheet of paper, make a final list of six examples of gravity at work. Make sure that at least two of these ideas are not described in this article. Write down notes about how gravity is involved in each of your examples. (You will use these notes later to do some writing.)

Make a poster that shows these six examples. Use markers to draw. Cut shapes from construction paper, or cut pictures from newspapers or magazines. Use stick-on labels to label each of your examples. Title your poster and write your name on the back.

Name _____

Questions about *Gravity*

Fill in the bubble that best completes each sentence.

1. Gravity is _____.

    ○ a pushing force that exists between all objects
    ○ a pulling force that exists only between large objects, such as planets
    ○ a pulling force that exists between all objects
    ○ a pushing force that exists only between large objects, such as planets

2. Objects with greater mass _____.

    ○ exert less gravity on each other
    ○ exert greater gravity on each other
    ○ are less likely to have gravity at all
    ○ are more likely to have gravity

3. Gravity between objects lessens _____.

    ○ when the objects are close to each other
    ○ as the objects move away from the Sun
    ○ as the objects move farther apart
    ○ during high tide

4. High and low ocean tides occur mainly because of _____.

    ○ the force of gravity between Earth and the Moon
    ○ cold water sinking to the bottom of the ocean
    ○ the force of gravity between the Sun and the Moon
    ○ hot air rising from the bottom of the ocean

5. If you went to the Moon, you could jump much higher and farther. This is because _____.

    ○ there is no air on the Moon
    ○ the Moon is far away from Earth
    ○ the Sun doesn't shine on one side of the Moon
    ○ the Moon has less mass and thus less gravitational pull

Vocabulary

Gravity

Write a word or phrase from the Word Box that correctly completes each sentence.

Word Box			
ebb	dense	bulges	flow
affects	rotates	orbit	solar system
mass	exists	exerts	

1. Our Sun and the planets and other objects that surround it make up our

   _____ _____.

2. Water in the ocean _____, or rises, on the side of Earth that is closest to the Moon.

3. Earth _____, or turns, all the time.

4. The movements of ocean tides are called _____ and

   _____.

5. Cold air is more _____ than warm air. This means that the molecules in cold air are packed more closely together.

6. Earth and the Moon _____, or apply, a pulling force on each other.

7. The planets _____, or move around, the Sun.

8. _____ is the amount of matter in an object.

9. Gravity _____ everyone on Earth. It has an effect on all of us.

10. Gravity occurs between you and your pencil and between you and your desk. It even

    _____ between people.

Gravity

Take out the notes you made earlier about gravity at work around you. Using your notes, explain how gravity is involved in each example on your poster.

Bouncing and Bending Beams of Light

Bouncing Beams of Light

Light travels in waves that we see as light rays. When light rays hit the surface of an object, the object absorbs some of them. Others are bounced back, or **reflected**. **Reflection** is the bouncing back of light rays from a surface.

Light rays are reflected back from a surface at the same angle they strike the surface. If the surface is rough, the reflected light bounces back in many different directions. If the surface is smooth, like a mirror, the rays bounce off the surface the same way they came to the surface. This is how a mirror image is formed.

Reflected light from a smooth surface is called **regular reflection**. Imagine the path of a bounced ball off a smooth surface to see how regular reflection works. A ball that is bounced straight down bounces straight up again. A ball that is thrown to the floor at an angle bounces up in the opposite direction, but at the same angle. That is how reflected light bounces in regular reflection.

You can study regular reflection by experimenting with mirrors. The image reflected from a mirror is reversed. Hold this page in front of a mirror. See how the words in the box appear backward?

> # I'm backward.

Now look at this box in the mirror. Can you read the message?

> # I'm not backward.

To see an image as it really appears, use two mirrors. Tape them together to form a corner.

Look into the corner your mirrors form. Your image looks right there. That's because it is reversed twice, once by the first mirror and once by the second. The second mirror reverses your image back to normal.

Probably most of the mirrors in your house are flat mirrors. But you may have seen curved mirrors in a funhouse at the fair. Curved mirrors change your image in funny ways. Some of the mirrors in a funhouse make you look short and wide. Others make you seem tall and thin. You can see how curved mirrors change your image by looking at yourself in a spoon. Look at your reflection in the spoon's handle. Look at it in the curved part of the spoon. Turn the spoon over and look at yourself in the hollowed-out part of your spoon.

Bending Beams of Light

Light rays always travel in straight lines. However, light rays change direction when they travel from one substance to another. This bending of light is called **refraction**. Refraction occurs because light travels at different speeds through different substances. For example, light travels slower through water than through air. The light rays reflecting off a coin at the bottom of a swimming pool are bent when they pass from the water to the air. As a result, the coin appears to be closer to the water's surface than it really is. Look down at your own body as you stand in a swimming pool sometime. Because of the way light bends when it moves from water to air, your underwater parts will look short and wide.

When light passes through water and glass, the bending of the light rays is even more obvious. Fill a glass with water. Place a ruler in the glass. Hold the glass so the water's surface is level with your eyes. Look at the ruler. It appears to be broken. That's because the light rays bend as they pass from the water to the glass, and then bend again as they pass from the glass to the air.

Experimenting with light rays is fun. You can make sentences look backward in a mirror, or rulers look broken in water. Try the experiments on the student activity page to see what else you can do with light.

Questions about
Bouncing and Bending Beams of Light

Fill in the bubble that best answers each question or completes each sentence.

1. Reflected light bounces back _____.

   ○ in many directions
   ○ at the same angle in which it hits an object
   ○ in a curved line
   ○ in circles

2. Why is your image not reversed when you look into the corner formed by two mirrors?

   ○ Images in mirrors are never reversed.
   ○ Light bounces back from a corner in many directions.
   ○ Your image is reversed twice, so it looks normal.
   ○ You can only see one mirror at a time.

3. What is the word scientists use for the bending of light?

   ○ refraction
   ○ reflection
   ○ direction
   ○ convection

4. Refraction makes a ruler in a cup of water look broken because _____.

   ○ water breaks plastic
   ○ the ruler is reflected at an angle
   ○ light rays bend when they move from water to air or air to water
   ○ the reflection of the ruler is broken into all the colors of the rainbow

5. Which of these statements is true about light traveling from one substance to another?

   ○ It does not bend.
   ○ It bends.
   ○ It scatters everywhere.
   ○ Light cannot travel from one substance to another.

6. In what way are reflection and refraction similar?

   ○ They both deal with sound waves.
   ○ They are both names for ways light bends.
   ○ They both happen when you look in a mirror.
   ○ They both deal with light rays.

Vocabulary

A. Write the correct word on each line below.

1. The bending of light is called _____.

2. When light bounces off an object, it is called _____.

3. _____ mirrors change your image in funny ways, making you look tall and thin or short and wide.

4. When light bounces off a smooth surface, the result is _____ reflection.

5. The image reflected from a mirror appears backward, or _____.

6. When reflected, light bounces back at the same _____ in which it hit an object.

B. Find and circle words in the puzzle that fit the clues given above.

A	C	F	G	H	I	M	O	P	N	S	T	C	K	I	J	M	N	O	B	A	T	O	D
D	E	C	O	L	L	E	P	R	E	N	A	C	R	C	O	J	M	B	O	T	R	O	M
B	N	O	E	S	S	U	J	E	O	T	A	P	E	P	C	E	L	E	S	T	A	M	A
N	O	G	I	G	L	S	T	I	A	N	T	E	V	E	T	U	N	G	A	J	E	U	S
B	O	M	J	U	N	H	T	M	N	W	X	Y	E	L	P	T	R	E	U	L	G	N	O
T	O	N	D	E	C	C	O	L	L	E	T	N	R	E	N	A	M	V	R	E	R	W	Z
Q	N	E	T	D	A	I	C	A	E	E	R	F	S	A	C	I	D	E	E	S	N	A	S
U	N	T	O	R	E	G	U	L	A	R	N	E	E	C	A	N	T	O	B	D	J	U	M
T	S	R	F	E	N	U	N	D	D	E	R	A	D	D	E	V	E	N	R	O	O	M	W
D	A	E	E	C	N	E	I	C	S	L	I	A	M	W	X	Y	Z	P	M	O	S	T	H
R	R	E	F	L	E	C	T	I	O	N	A	V	E	L	C	S	A	N	G	L	E	L	V
I	P	E	P	S	R	Q	U	V	M	W	T	U	I	S	M	N	S	B	T	O	D	S	H

Name _____

Reflection and Refraction Experiments

Try the following experiments. In the boxes provided, draw pictures of what you see. Then complete the sentence below each box.

Experiment 1

- **Materials:** mirror, flashlight

- **Steps:** Place a mirror on the floor of a dark room. Shine the flashlight into the mirror. Look up at the ceiling above the mirror. Draw what you see in the box.

I saw the light on the ceiling because the mirror _____ the light from the flashlight.

Experiment 2

- **Materials:** drinking glass, water, coin

- **Steps:** Place the coin at the bottom of the glass. Pour water slowly into the glass until you see something strange. Draw what you see in the box.

It looks like there are two coins in the glass because light rays are _____ as they move through the water and the glass.

Experiment 3

- **Materials:** drinking glass, water

- **Steps:** Fill the glass with water almost to the brim. Look down into your glass. Draw what you see in the box.

The part of the table directly under your glass looks _____ than the rest of the table because light rays _____ off the bottom of the glass are _____ at the water's surface.

A Better Gizmo

"Congratulations! You have been accepted into the Incredible Gizmo School of Inventing. Our motto is "The more parts, the better!" I'll show you what I mean with our first slide. Will someone get the lights?"

The auditorium fell dark, and a goofy-looking contraption appeared on the screen. As the students studied the slide, there was a ripple of laughter from the back row. "Who would ever want that?" shouted a young girl. "It's silly."

"Well, so what?" said the speaker. "Inventions can be silly."

"Not the good ones," said the young girl. "The best inventions are practical. They serve a purpose or fulfill a need. They try to simplify a task. Pocket calculators, for instance. Computers." She pointed to the screen. "That invention does just the opposite. And besides, it's much too complicated. An invention should never have parts it doesn't need." And before anybody could stop her, the young girl was on her feet and heading for the stage.

"Suppose you have a great idea," the young girl said to the audience. "Maybe you know how something can be improved. Take the phonograph, for instance. The first one was built by Thomas Edison. He recorded sounds on wax cylinders. Then, to play the sounds, he ran a needle along grooves in the wax. This produced **vibrations**, which were then **amplified**—or boosted—by a horn. But the grooves were uneven, so the sound quality wasn't very good. Then Emile Berliner built what he called the **gramophone**. This was an improvement because it could play flat records whose grooves were all the same depth, but there was a drawback. The gramophone was driven by a spring motor that had to be wound up with a hand crank, so you had to rewind the gramophone to keep it going. Well, the next improvement was a record player that ran on electricity. Get the picture?"

The audience nodded, and a man in the front row began to take notes.

"Who is this kid?" whispered a woman in a large hat.

"All right," said the young girl. "Your great idea is up here." She tapped her head. "But what you can imagine may not necessarily work. So the first thing you have to do is draw a **sketch** of your invention.

"In your sketch, label all the parts of your invention and show how they fit together. Will you need springs? Screws? Washers? Clamps? Hinges? Hoses? Pipes?"

By now just about everybody in the audience was taking notes, including the speaker from the Incredible Gizmo School of Inventing.

"When you're satisfied with your design, you can start building a **breadboard**. Inventors make their breadboards out of whatever parts they have handy. It's not a finished product, just a rough model to prove that your invention will actually work. The best inventions use the fewest parts needed to get the job done. As you can see, this invention has some extra parts."

The audience burst out laughing.

"At this stage, one of three things may happen.

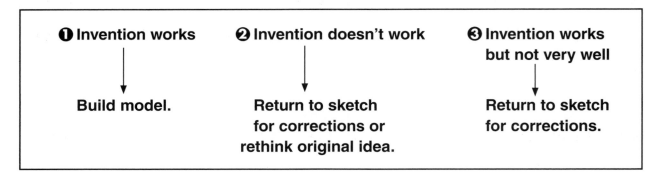

❶ Invention works	❷ Invention doesn't work	❸ Invention works but not very well
↓	↓	↓
Build model.	Return to sketch for corrections or rethink original idea.	Return to sketch for corrections.

"If your invention works, you can build your working **model**, but you have to consider several things: Who is going to use it? Adults? Children? Will it be used a lot or only once in a while? Will it be used outside or indoors? Should it be waterproof? Shatterproof? Lightweight?"

"What about cost?" asked a man.

"That's a good question," said the young girl. "How much will it cost to build your invention? If you use very expensive materials, the cost might be so high that nobody will want to buy it. This is when you may decide to use tin instead of silver, or balsa wood instead of maple.

"Once all these details are ironed out, it's time for your **prototype**. A prototype of your invention is a one-of-a-kind, handmade sample that looks and works exactly the same as if it came out of a factory. It has to be beautiful," said the young girl, reaching into the pocket of her overalls. "It has to be what everybody wants, what everybody has been hoping for. Like this," she said, holding up a strange-looking gizmo.

"Wow!" someone cried. "What is that?"

"This?" said the young girl. "It's a clock that never loses time. It's a light bulb that never burns out. A pen that corrects your spelling. A pillow that's always fluffy. A computer chip with infinite memory. It's whatever fabulous idea you can think of. It's your invention."

Questions about
A Better Gizmo

Answer the following questions in complete sentences.

1. Why isn't the invention pictured in the story a particularly good one? List two or more reasons.

   _____

   _____

   _____

2. Fill in the missing words or phrases in this description of Thomas Edison's phonograph.

   Sounds are recorded on _____. A _____ runs

   along grooves in wax and _____ are produced which are

   _____ by a horn.

3. Why is it important to start with a sketch of your invention before you begin building it?

   _____

   _____

4. What is the difference between a working model of your invention and a prototype?

   _____

   _____

5. Jason has what he thinks is a fantastic idea for an invention. In fact, he's so sure of himself, he's decided to skip building a breadboard and go right to his model. Why is this unwise?

   _____

   _____

Vocabulary

Name That Product!

When a product or an invention is the first of its kind, sometimes people start calling the product by its brand name. Frigidaire, a name that came to mean "refrigerator," made one of the earliest refrigerators. As you can imagine, companies love when this happens.

Here are other trade names that have informally entered our language as words. What do they really stand for?

1. Kleenex   _____

2. Jell-O   _____

3. Scotch tape   _____

4. Xeroxing   _____

5. Band-Aid   _____

6. Popsicle   _____

7. Rollerblades   _____

8. Q-Tips   _____

Gizmos Unlimited

Gizmo, *doodad*, *thingamajig*, and *whatnot* are all words that stand for the same thing, but what exactly is that thing? Write your definition for a *gizmo* below.

giz' mo (noun)

_____

_____

_____

_____

Name _____

A Better Gizmo

As a student of the Incredible Gizmo School of Inventing, you are required to build an invention. Your objective is to make a very simple task as complicated as possible.

Choose a simple device or invention that already exists (like a pencil sharpener) and think of a way to make it complicated. Your invention should have at least 10 parts. Draw a sketch of your invention and label it. The labels should explain what each part does and how your device works.

All Rise

What Is Yeast?

Yeast is a living **organism**, yet it is not a plant or an animal. It is a **fungus**. Fungi do not make babies or lay eggs like animals. They do not sprout from seeds like most plants. Most types of yeast reproduce through **budding**. In budding, a part of one yeast cell pinches off to make a new cell.

Fungus was once considered a simple plant. But it does not make its own food like plants. Fungi have no **chlorophyll** to make food. Instead, they absorb food from other substances.

Yeast is a very small, one-celled fungus. Yeast cells look like tiny eggs. They can only be seen under a microscope. One hundred billion yeast cells weigh less than an ounce.

Wild yeast is present in the air. It also lives in the soil and on plants. The yeast that is important to you is grown in factories. This commercial yeast is used in making medicines, baked goods, and drinks.

Who Discovered Yeast?

Five thousand years ago, Egyptians discovered that bread dough would rise if left out for a long time. They did not know wild yeast from the air was getting into their dough. They just knew the bubbly batter baked into a light and airy loaf of bread. Ancient Egyptians also knew fruits and grains could turn into wine and vinegar. They may have thought magic was responsible for the changes. Since they had no microscope, they could not see yeast. They had no way of knowing live fungi grew in their bread dough and fruit juices.

Louis Pasteur

Louis Pasteur was the first person to prove that living organisms caused grape juice to turn to wine. In 1847 he conducted experiments that proved yeast from the air entered the juice. He showed that **enzymes** (proteins found in living organisms) in the yeast broke down sugars and starches, turning them into **alcohol** and a gas called **carbon dioxide**. The process Pasteur demonstrated is called **fermentation**.

How Does Bread Rise?

Bread dough is made of flour, water, butter, salt, sugar, and yeast. The water and flour combine into a gluey mixture called **gluten**. The yeast ferments the sugar. Gluten traps the carbon dioxide bubbles made during fermentation. Butter makes the gluten slippery so the bubbles can expand more easily. When the trapped bubbles grow, the bread dough rises. Salt makes the bread rise at the correct speed and helps the rising dough stay strong.

How Is Yeast Used?

Commercial yeast is sold in many forms. Fresh, wet yeast is made into a smooth paste and pressed. Then the yeast blocks are sold to bakers. Sometimes it is sold all crumbled up in big bags. Wet yeast is used to make baked goods rise.

Moisture is taken out of wet yeast to make semi-dormant **dry yeast**. Warm liquid must be added to dry yeast to make it active again. Dry yeast is sold in grocery stores in jars and packages. It is also used to make baked goods rise. If you want to make bread at home, you would buy a package of dry yeast.

Yeast is also used to ferment grains and fruits in the making of alcoholic drinks. The fermented product becomes the drink. The yeast itself does not stay in the drink. It is either reused to ferment more grains or fruits, discarded, or added to animal feed.

Nutritional yeast is another type of yeast. Nutritional yeast does not make dough rise. It is used to make some kinds of antibiotics. It is also a good source of B-complex vitamins. Nutritional yeast is sometimes fortified with other vitamins and proteins, and sold in health food stores as a pill or a powder. **Brewer's yeast**, one type of nutritional yeast, is a by-product of making beer. It is used to flavor foods, mixed into animal feed, and sold in health food stores for its health benefits.

Humans have been baking breads and making wines for thousands of years. Now that we know how yeast works, it has become more valuable than ever.

All Rise

Questions about *All Rise*

Fill in the bubble that best answers each question.

1. Why do you think this article was entitled *All Rise*?

   ○ Yeast makes bread dough rise.

   ○ The article is so important that you should stand up when you read it.

   ○ Yeast was first discovered by a judge.

   ○ The article is about flying objects.

2. Which phrase best describes *All Rise*?

   ○ *All Rise* is a short story based on the history of yeast.

   ○ *All Rise* tells what yeast is, how it was discovered, and how it is used today.

   ○ *All Rise* tells readers about many kinds of fungi.

   ○ *All Rise* is a biography about Louis Pasteur.

3. How does yeast get food?

   ○ Yeast eats food with teeth like many animals.

   ○ Yeast makes its own food like a plant.

   ○ Yeast doesn't need food.

   ○ Yeast absorbs food from other substances.

4. Which statement about Louis Pasteur is **not** true?

   ○ Louis Pasteur proved yeast was a live organism.

   ○ Louis Pasteur demonstrated the process of fermentation.

   ○ Louis Pasteur baked the first loaf of bread.

   ○ Louis Pasteur showed that yeast breaks down sugars and starches.

5. Which statement about the uses of yeast is true?

   ○ Commercial yeast is always used to make baked goods rise.

   ○ Brewer's yeast is used to make bread.

   ○ Wet and dry yeasts are used in baking.

   ○ The final product of wine contains a lot of yeast.

Vocabulary

Use a dictionary or context clues from the article to help you match the vocabulary words from *All Rise* with their definitions.

Word Box			
fungus	chlorophyll	reproduce	absorb
budding	enzyme	semi-dormant	substance
carbon dioxide	fermentation	organism	fortify

_____ 1. to take in like a sponge

_____ 2. the physical matter of a thing, material

_____ 3. to add vitamins to something to increase its nutritional value

_____ 4. a protein found in living organisms that causes change in a substance

_____ 5. the breakdown of a compound such as when yeast absorbs sugars and starches to produce carbon dioxide and alcohol

_____ 6. a living organism that is neither a plant nor an animal

_____ 7. to bear offspring or produce again

_____ 8. a colorless, odorless gas

_____ 9. the substance in plants that assists in the making of food

_____ 10. any living thing

_____ 11. temporarily inactive

_____ 12. the pinching off of part of a cell to form a new cell

Name _____

The Work of Louis Pasteur

Louis Pasteur

Louis Pasteur's research on fermentation was followed by much more research. Read the summary of the works of Pasteur below. Then decide which works saved lives and which saved businesses. Write your answers on the lines provided.

Louis Pasteur (1822–1895) was a very busy chemist. His research on fermentation helped wine makers make better wine. His understanding of germs helped people stay safe. He showed that heating milk, honey, and wine killed germs. He said that sterilizing medical instruments and washing hands before surgeries in hospitals killed germs too. Pasteur created vaccines to prevent the spread of disease in animals. He helped save the silk industry by stopping a disease in silkworms. He made vaccines for diseases that struck sheep, chickens, and pigs too. Then Pasteur discovered three bacteria that cause human illness. Finally, he developed a treatment for rabies. He tested his rabies treatment on a ten-year-old boy who had no other chance of survival. The boy lived. Pasteur opened a clinic for rabies treatment. He worked there until his death.

Works That Saved People's Lives

_____

_____

_____

_____

Works That Saved People's Businesses

_____

_____

_____

Making Your Car Move

The first car ever built was powered by steam. A large boiler in the back of the car was filled with water and heated. Steam built up pressure that drove the car's engine. Yet water needed to be added to steam cars often. They slowed down as their steam pressure decreased with time. So steam cars lost their market. Today only members of special clubs drive steam-powered cars.

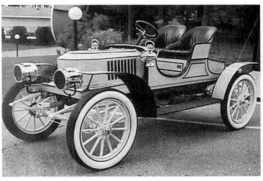

Most modern cars are powered by gasoline. But cars that run on gasoline pollute the air. Also, the world's supply of **petroleum** (the material used to make gasoline) is limited. That's why scientists today are working on **alternative** ways to make cars move.

In the early days of cars, it was uncertain how cars should be powered. Makers of steam, electric, and gas-run cars raced their vehicles to prove their worth. The Stanley Steamer became a popular car when it won many such races. This photograph is of a 1908 Stanley Steamer 20 hp H-5 Gentleman's Speedy Roadster.

Electric Cars

A **battery** powers an electric car. When the battery runs down, it is **recharged** from an outlet. Power plants that send energy to the outlet make most of their electricity by burning coal. So electric cars do not use **renewable energy sources**. Still, they have two **environmental** advantages over gas-powered cars:

1. Electric cars are very efficient. They convert up to 70 percent of their energy into forward motion. Gasoline-powered cars are no more than 25 percent efficient. The rest of their energy is lost as heat. So electric cars use up less of our **natural resources**.

2. Electric cars by themselves cause no **pollution**. And the power sources that charge them follow strict laws that limit their pollution too. A city full of electric cars would be **smog-free**.

Still, electric cars have problems. They cost a lot, and they are not as practical as gas-powered cars. An electric car's battery needs to be recharged every 100 miles or so. If the car's driver turns on the radio or lights, the car needs to be recharged even sooner. The battery takes at least three hours to recharge. So the electric car can't make long trips.

Read and Understand, Science • Grades 4–6 • EMC 3305

Solar-powered Cars

One way to assure an electric car does not add to the pollution problem is to recharge its battery with sun power. Students in high school and college science classes build and race **solar-powered** cars today. These cars have solar panels that collect energy from the sun. The energy is transferred to a battery, like the one on an electric car. Since solar cars use a renewable energy source, they do not create pollution.

M-Pulse
The 2001 University of Michigan Solar Car Team

Yet the solar car has problems. Like the electric car, the solar car's battery needs to be recharged often. It takes at least three hours to recharge in bright sunlight. On a cloudy day, the battery cannot be recharged at all. For these reasons, solar cars are not used for everyday driving.

HEVs

Some major car companies have come out with **hybrid electric vehicles**. HEVs are a cross between electric and gas-powered cars. An HEV uses half the gas of a normal car. It also creates less pollution. Unlike the electric car, the HEV can make long trips because it can run on gasoline too. Still, the HEV does not make use of a renewable energy source.

Biomass Fuels

All plant material contains energy from the sun. **Biomass** is plant or animal material that contains stored energy. Biomass such as corn can be used to make alcohol, a liquid fuel. This alcohol can be mixed with gasoline to produce **gasohol**. Gasohol is used a lot in some countries. It creates less pollution than gasoline. It also wastes less petroleum. But it does have its problems. Gasohol costs a lot to make because it takes a lot of corn to make a small amount of gasohol. A car running on gasohol is able to travel fewer miles on the same amount of fuel as one running on gasoline. There is some concern that gasohol may damage a car's fuel system.

No car powered by an alternative energy source is without its problems. But carmakers are addressing drivers' concerns. In the near future, your family's car may very well run on something other than gasoline.

Name _____

Questions about
Making Your Car Move

Answer the following questions in complete sentences.

1. What energy source powered the first car ever built?

_____

2. Which of the energy sources mentioned is renewable?

_____

_____

3. In what types of places would a solar-powered car be most practical?

_____

_____

4. Which renewable energy source mentioned in the story can run a traditional car?

_____

5. How are the solar-powered and the electric car alike? How are they different?

_____

_____

_____

6. Which cars are not good for long-distance driving and why?

_____

_____

7. How do you think the problem of powering cars will be solved?

_____

_____

_____

_____

Making Your Car Move

Vocabulary

Read the glossary below to review the meanings of some of the words from the story you just read. Then complete the sentences that follow.

Glossary	
alternative energy	energy produced from a source not commonly used
electricity	a form of energy that can be easily moved from one place to another through wires
hybrid	a cross between two types or groupings
renewable energy	energy produced from natural resources that can be replaced
solar power	energy from the sun that is turned into electrical energy

1. A _____ electric car is a cross between an electric and a gas-powered car.

2. Wind can be used to produce power. Wind power is both an _____

   _____ source and a _____

   _____ source.

3. Solar panels and batteries are used to operate cars that run on _____

   _____.

4. Electric cars make use of _____ produced by power plants.

Name _____

Weighing the Options

Driving a car that is powered by a renewable energy source is one way to cut down on pollution and conserve gas. There are other ways too. Write the pros and cons of using the transportation methods listed below on the lines provided.

Take a Bus or Subway

Advantages _____

_____

_____

Disadvantages _____

_____

_____

Ride a Bicycle

Advantages _____

_____

_____

Disadvantages _____

_____

_____

Walk

Advantages _____

_____

_____

Disadvantages _____

_____

_____

Comets

"I'm supposed to write a report about comets," said Jan. "It's for science."

"Supposed to?" asked Mom. She looked up from her computer.

Jan said, "I did the research and everything, but I don't know what to write. I keep putting words on paper, but they're not very good. I've been working on it for hours."

"What have you written?" asked Mom. She leaned back in her chair to listen.

Jan picked up her paper and read. "A comet is a huge ball of ice and rock. The center, or nucleus, is wrapped in a layer of dust and vapor called the coma. The thing most people notice first when they look at a comet is the tail. When a comet is near the sun, part of it melts. Bits of ice and rock turn to vapor and dust. The solar wind blows the dust and vapor away from the comet, forming a tail. This is why the tails of comets always point away from the sun.

"As a comet approaches the sun, the coma and tail reflect the light. For a few weeks, we see a tiny smudge of light in the night sky. In fact, the tails of many comets are millions of kilometers long.

"Comets move in oval-shaped orbits around the sun. Some take millions of years to make their orbits. Others take as little as three years. The well-known Halley's comet takes between 74 and 79 years to orbit the sun."

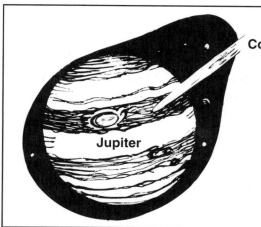

Comet

Jupiter

The orbits of comets change over time. A small part of every comet burns away when it comes close to the sun, changing its orbit slightly. Sometimes, when a comet comes close to a planet, the planet's gravity changes its orbit. In 1992 comet Shoemaker-Levy was found to be in a very close orbit around Jupiter. Over the next couple of years, it broke up and slammed into the planet.

When Jan stopped reading, her mother said, "Sounds pretty good, so far. By the way, I saw Halley's comet in 1986." She looked at her daughter and tilted her head. "Think of it, Jan. I was twenty years old when I saw Halley's comet. I'll be in my nineties the next time I see it."

Jan did some quick figuring. "You'll be 95," she said. "I read that Halley's comet will be back in 2061."

"Ninety-five years old!" whispered Jan's mother. She shook her head. "It's hard to imagine what my life will be like then."

Then, something startling occurred to Jan. "I'll be 69!" she said. "I could be a grandmother!" The two of them laughed at the idea.

"You've seen a comet too," said her mother. "When you were five years old, we all went out to see comet Hale-Bopp. It was as if a giant had picked up a paintbrush and dipped it in light, then touched it against the sky."

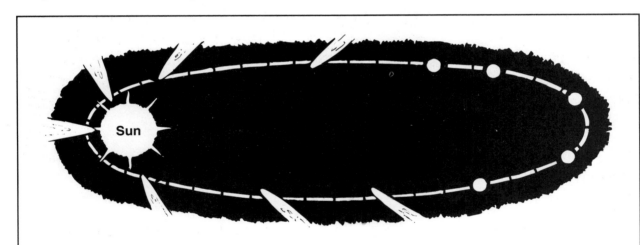

Astronomers describe Hale-Bopp as a long-period comet. Long-period comets take 200 years or more to orbit the sun. Halley's comet is a short-period comet, meaning its orbit takes less than 200 years.

Jan didn't remember seeing comet Hale-Bopp, but she had read about it. It shone in Earth's sky in the spring of 1997. It was especially bright, because its coma was wide and reflected a lot of light. "It's a long-period comet," she said, checking her notes. "It will be back in 2,380 years."

"Oh, my," said Mom. "I guess I won't be around to see it."

"Nope, neither will I." Jan sighed. "But hopefully, I will have finished my report."

Questions about *Comets*

Fill in the bubble that best completes each sentence.

1. The nucleus of a comet is made of _____.

   ○ ice and vapor
   ○ ice and rock
   ○ vapor and light
   ○ rock and vapor

2. As the comet approaches the sun, the coma and tail _____.

   ○ reflect the sun's light
   ○ burn up completely
   ○ spread out, becoming much larger
   ○ become difficult to see

3. The tail of a comet always points away from the sun because _____.

   ○ the comet only moves directly toward the sun
   ○ we cannot see part of the tail that is closest to the sun
   ○ the solar wind blows dust and vapor away from the comet
   ○ waterdrops freeze on the cooler side of the comet, becoming visible

4. Comets generally move in _____.

   ○ oval-shaped orbits
   ○ large groups
   ○ straight lines across the solar system
   ○ winter only

5. According to the story, Halley's comet _____.

   ○ orbits Earth
   ○ takes more than 200 years to make its orbit
   ○ has slammed into a planet
   ○ takes less than 200 years to make its orbit

Vocabulary

A. Draw a comet in orbit around the sun. Label the comet's nucleus, coma, and tail.

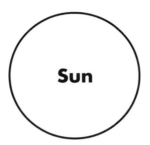

Sun

B. Write the base word for each of the words below. Then write a different word formed from the same base word. Finally, write a sentence using the new word. The first one has been done for you.

1. **startling**   base word: _startle_   new word: _startled_

   You startled me when you spoke so suddenly.

2. **approaches**   base word: _____   new word: _____

   _____

3. **reflected**   base word: _____   new word: _____

   _____

4. **tilted**   base word: _____   new word: _____

   _____

5. **figuring**   base word: _____   new word: _____

   _____

Using Time Lines

Time lines show the order in which events occur, both in fiction and in nonfiction. Time lines are especially helpful in understanding time relationships between several events.

A. Use clues in the story to help you complete the time line below. It shows events surrounding Halley's comet, comet Hale-Bopp, and comet Shoemaker-Levy.

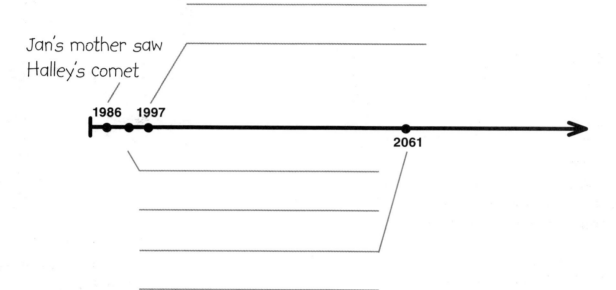

Jan's mother saw
Halley's comet

1986 1997

2061

Bonus Question: In what year was Jan born? Mark it on the time line above.

B. Now use the time line below to show at least four events that have happened this school year. Mark a dot on the time line for each event. Write when the event occurred below the dot. Write a brief description of the event above the time line.

Asking Mr. Aims

Kate was studying careers in school. Each of the students in Kate's class was asked to talk to one adult about his or her job. Kate talked to Mr. Aims.

Mr. Aims was Kate's neighbor. He worked at a rock quarry. Kate had been to the quarry so she knew it was a big pit of rock and stone. She also knew large trucks and machines parked at the quarry, but she didn't know what Mr. Aims did at the pit.

Kate: Good morning, Mr. Aims. Thank you for talking to me about your job. I know you work at the rock quarry on the edge of town. What do you do there?

Mr. Aims: Well, Kate, the men and women at the rock quarry break large rocks into smaller blocks and stones. The blocks and stones are used to make building materials. My job is to blast big slabs of limestone with dynamite. My friend Tom scoops the stone up with a big backhoe. My friend Jan drives a big truck full of the limestone pieces to a factory where the stones are turned into glass, steel, and cement.

Kate: You mean shiny silver steel is really made of rock?

Mr. Aims: Steel is made when iron ore (rock that contains iron), carbon, and limestone are heated in a very hot furnace. The chalk your teacher uses in the classroom is also made of limestone. Limestone is most useful when it is heated to a high temperature and turned into lime. Lime is used in the making of cement, paper, glass, water softeners, antacids, and soil additives. Lime is even used to refine sugar and tan leather.

Kate: Are other rocks as useful as limestone?

Mr. Aims: Artists carve marble into sculptures. Slate makes shingles for rooftops. Jewelry is made from quartz. Coal is used as fuel. Granite is used to construct buildings.

Kate: There are so many different kinds of rocks. What makes one rock different from another?

Mr. Aims: Different kinds of rocks are formed in different ways. Some rocks come from inside the Earth. When a volcano erupts, hot liquid rock called magma bursts out of the Earth as lava. When the lava cools, it forms hard rocks. Lava that cools quickly becomes a shiny black rock called obsidian. Lava that cools before gases inside it can escape forms pumice. Pumice is so light and full of holes that it floats in water. Sometimes magma hardens underground instead of shooting out of a mountain as lava. Granite is a rock that is formed underground when magma cools slowly. Rocks that form from magma are called igneous rocks. Igneous means "made from fire."

Kate: Do all rocks come from inside the Earth?

Mr. Aims: No, some rocks are formed on Earth's surface. Sedimentary rocks are made from mud and tiny pieces of sand and shells that get pressed together on the ocean floor. The limestone that I work with at the quarry is made of animal shells.

Kate: I know that shells pile up on the ocean floor when sea animals die, but where does sand come from?

Mr. Aims: When wind, ice, and rain wear down big rocks all around the world, pieces of the rocks crumble and fall into streams. Slowly, the rock pieces make their way to oceans. As they tumble along, the rocks break into even smaller pieces. The edges of the small pieces of rock are smoothed by water and wear. Slowly, the weathered, crumbled pieces of rock break into bits of sand. Layers of sand and mud build up on the ocean floor as more and more rocks bump their way down streams to the ocean. Lower layers of mud and sand on the ocean floor get pushed together so tightly that they form hard rocks.

Kate: So sedimentary rocks are layered rocks?

Mr. Aims: Yes, sedimentary rocks are layered rocks of shell, mud, and sand formed on the ocean floor. Igneous rocks are made from magma from below Earth's surface. And there is one other kind of rock—metamorphic rock. Metamorphic rocks are made from the two other kinds of rocks. Heat and pressure from inside the Earth change one kind of rock into another kind of rock. Limestone becomes marble when it is heated and pressed. Shale turns into slate, and sandstone turns into quartzite.

Kate: You know a lot about rock, Mr. Aims. Thank you for talking to me about your job and about all the uses of the different kinds of rocks.

Rocks Times Three

There are three types of rocks:

Sedimentary *rocks are formed when particles of matter are laid down in many layers and pressed together.*

Igneous *rocks are formed when hot liquid rock from deep inside the Earth comes to the surface, cools, and hardens.*

Metamorphic *rocks are formed when heat and pressure inside the Earth change one kind of rock into another kind of rock.*

Questions about
Asking Mr. Aims

Fill in the bubble that best answers each question.

1. Limestone is an example of which kind of rock?

   ◯ metamorphic rock
   ◯ igneous rock
   ◯ sedimentary rock
   ◯ molten rock

2. Which statement about sedimentary rock is **not** true?

   ◯ Sedimentary rock means "rock made from fire."
   ◯ Sedimentary rock is layered.
   ◯ Sedimentary rock is formed on the ocean floor.
   ◯ Sedimentary rock is made from packed sand, mud, and/or shells.

3. Which statement about metamorphic rock is true?

   ◯ Sandstone is a metamorphic rock.
   ◯ Metamorphic rock is cooled lava.
   ◯ Metamorphic rock is made from the two other kinds of rocks.
   ◯ Metamorphic rocks float because they are full of trapped gases and holes.

4. Which force does not break down rocks in nature?

   ◯ wind
   ◯ ice
   ◯ water
   ◯ sunlight

5. Which is **not** a use of lime?

   ◯ It is used in the process of refining sugar.
   ◯ It is used to make lime-flavored drinks.
   ◯ It is present in cement.
   ◯ It is an ingredient in glass.

6. What does the word *igneous* mean?

   ◯ made from pressure
   ◯ made from other rocks
   ◯ made from sand
   ◯ made from fire

Vocabulary

Decide whether each of the statements about the words from the story are true or false. Circle **T** if the statement is true. Circle **F** if the statement is false. If the statement is false, rewrite it to make it true.

1. **T**   **F**   Limestone is broken down in quarries and used to make building materials and other products.

   _____

2. **T**   **F**   Slate, which is used to make shingles for rooftops, is a sedimentary rock.

   _____

3. **T**   **F**   Obsidian is a shiny black rock.

   _____

4. **T**   **F**   Granite is a metamorphic rock.

   _____

5. **T**   **F**   Pumice floats in water.

   _____

6. **T**   **F**   Rocks formed from molten rock are igneous rock.

   _____

7. **T**   **F**   When tiny pieces of shell get pressed together, they form metamorphic rocks.

   _____

8. **T**   **F**   Heat and pressure from inside the earth form metamorphic rock.

   _____

9. **T**   **F**   When limestone is heated, it becomes sandstone.

   _____

10. **T**   **F**   Glass jars are made from quartz.

   _____

Rock Detective

Identify the rocks pictured here by reading their descriptions below.

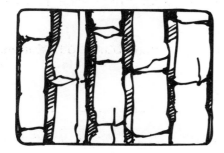

1. _____

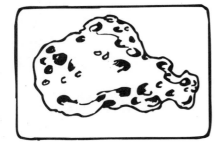

2. _____

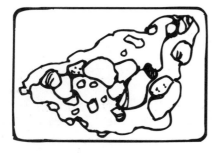

3. _____

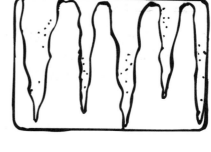

4. _____

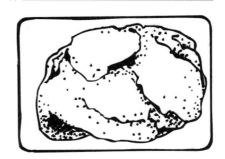

5. _____

6. _____

Basalt: This igneous rock often appears in columns.

Limestone: In addition to being found in rock quarries, this rock often hangs like icicles from the ceilings of caves.

Conglomerate: Pebbles are packed with mud and sand to make this rock.

Pumice: This lighter-than-water igneous rock is full of holes.

Obsidian: This shiny black rock is formed when lava cools quickly.

Sandstone: The layers in this sedimentary rock are easy to see.

The Greatest Show on Earth

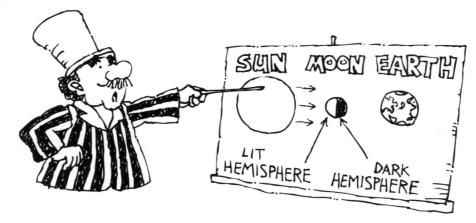

"Hurry, hurry, hurry! Step right up, folks. Step right up."

Nathan felt a rush of excitement shoot through his body. It was as if someone had filled all his veins and arteries with soda pop.

"You are about to see," said the colorful man in the red-and-white striped suit, "a most astonishing sight, a sight that will take your breath away!"

Nathan's heart was pounding. He was hardly breathing already.

"Ladies and gentlemen!" cried the man, pointing to the horizon. "May I present the star of our show!"

All heads turned to see a newly risen September sun perched just above the far hills.

"Yes, ladies and gentlemen, you are looking at the giant ball of hot gas around which Earth and all the other planets move!"

The crowd oooed.

"But wait," said the man, "because also in the sky, and moving steadily toward the spot where it will make history, is the Moon! Now, I know you can't see it, but trust me, it's there," and all at once there was a little puff of smoke, and a fabulous chart appeared.

Nathan knew that the chart showed the phase called new moon. A new moon occurs when the Moon slips in between Earth and the Sun. As it passes by, the Sun's light shines only on the hemisphere that is facing it, so the hemisphere facing Earth remains in the dark.

The man in the wonderful striped suit tapped the chart with his finger. "Everything is on the move," he said. "The Moon orbits Earth while Earth and the Moon, like partners in a graceful waltz, orbit the Sun. It is a beautiful dance in the heavens, but behold! What you are about to see is the most spectacular part of the entire dance. What you are about to see is a total eclipse of the Sun!"

The crowd gasped.

"The Sun and the Moon are different sizes, but because the Moon is so much closer to us, it appears to be about the same size as the Sun. So when the Moon is in just the right place, it can cover up the Sun! 'Impossible,' you may cry, but it's true, and I will prove it to you, but first I will need a volunteer."

Quick as a flash, Nathan's hand shot up.

"Ah," said the man, grinning widely. "The little boy in green. Step forward, please," and as Nathan climbed onto the stage, the man handed him a grape. "You," he said to Nathan, "are Earth." Nathan looked out at the crowd and giggled. "The grape I have just given you is the Moon, and this," he said, holding up a large orange, "is the Sun. They're different sizes, aren't they? You wouldn't think that your tiny grape could completely cover up the orange, would you? But I will show you how it's done. Stand over there, please, and hold the grape up in front of your face. I'll hold up the orange. Now, I want you to shut one eye and move the grape until it lines up exactly in front of the orange."

Nathan did as he was told.

"Got it?" asked the man, and Nathan nodded. "All right, slowly move the grape closer to your face."

Every eye was glued to Nathan and the man in the star-spangled suit.

"Wow!" shouted Nathan. "It worked! I covered the orange with my grape!"

And that's what happens during a total eclipse of the Sun.

But the Moon can't hide all of the Sun. It can't hide the outermost part of the Sun's fiery atmosphere, and during a total eclipse, you can see it. The edge of the atmosphere is called the Sun's corona.

But now the show was about to start, and the man quickly handed out little filters to protect everyone's eyes.

And through the filters they watched in silent wonderment as ever so slowly the Moon moved into **position**, taking bigger and bigger bites out of the Sun. Nathan could feel it growing cooler and cooler during these **partial phases** of the eclipse.

And then suddenly all the birds fell silent, believing that night had come.

"Now!" shouted the man in the red-and-white star-spangled suit. "Remove your filters! We have totality!"

Nathan tore off his filter and gasped.

The Sun was completely hidden, its golden corona rippling and shimmering like a fiery ring around the great black disk that was the Moon. Nathan knew there would only be a few minutes of totality because the Moon and Earth keep moving, so he watched as hard as he could. Yes, he thought, as a brilliant burst of sunlight signaled the end of totality, a total eclipse of the Sun really is the greatest show on Earth.

During a solar eclipse the Moon progressively covers more and more of the Sun. At totality, only the Sun's corona is visible.

Questions about
The Greatest Show on Earth

Fill in the bubble that best answers each question.

1. During which phase can a total eclipse of the Sun occur?

   ○ crescent
   ○ full moon
   ○ half moon
   ○ new moon

2. What happens during a total eclipse of the Sun?

   ○ The Moon passes between the Sun and Earth.
   ○ Earth passes between the Moon and the Sun.
   ○ The Sun passes between Earth and the Moon.
   ○ The Moon passes between Earth and Venus.

3. When Nathan got home, he wanted to show his sister Amy how a total eclipse occurs. He found grapes, but Amy had eaten the last orange. What could Nathan use instead of the orange to represent the Sun?

   ○ a raisin
   ○ a banana
   ○ an apple
   ○ an apple seed

4. When is the only safe time to look at the Sun?

   ○ during the partial phases of an eclipse
   ○ during totality
   ○ in the early morning
   ○ in the late afternoon

5. How is it that the Moon can block out the Sun?

   ○ The Moon is nearly as big as the Sun.
   ○ The Sun looks smaller at certain times of the year.
   ○ The Moon seems to take bites out of the Sun until it is gone.
   ○ The Moon appears to be the same size as the Sun because it is closer to Earth.

Vocabulary

Prefixes are terrific little word parts. If you know what different prefixes mean, you can unlock the definitions of all kinds of words. For example: the prefix *hemi-* means "half." So which of these is a **hemi**sphere?

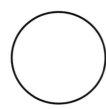 

A. Here are some commonly used prefixes with their meanings. For each word below, circle the prefix and write the meaning of the word on the line.

bi–two	**un**–not	**semi**–half

1. unhurried   _____

2. semicircle   _____

3. unripe   _____

4. bicoastal   _____

5. semidarkness   _____

6. bicolored   _____

7. bipolar   _____

8. unseasonable   _____

B. Which word from the list above best describes each of the following?

1. Jeffrey's yellow-and-green striped shirt   _____

2. snow in June   _____

3. the tortoise who ran against the hare in their famous race   _____

4. what totality creates   _____

5. this: ⌒   _____

6. a flight from here to there   _____

7. a magnet   _____

8. a green banana   _____

The Big Scoop

Wow! Lucky you! *The Daily Star and Planet* has agreed to publish your eyewitness account of the total solar eclipse you just witnessed, so you'd better hurry up and write it.

Describe the eclipse from beginning to end. Include how you felt, as well as what you and others saw. You may even want to interview some of the other people who saw the eclipse with you: Nathan, people in the crowd, and even the man in the star-spangled suit. If you quote someone, be sure to use quotation marks.

Remember to give your news story a catchy title.

_____

_____

_____

_____

_____

_____

_____

_____

_____

_____

_____

_____

_____

_____

_____

_____

_____

_____

_____

A Compound Mystery

Osgood Blastwood lay on the floor of his bathroom, a purple bucket beside him, a pair of green rubber gloves on his hands. He was stone-cold dead.

Inspector Yoghurty shook his head. "Poor guy," he said, "he didn't even get to finish washing out the tub."

"What do you think killed him, Chief?" asked Deputy Sikes.

"Don't know," said Yoghurty, rubbing his chin.

"Ah," said a voice. "But I do!"

Heads turned. "Claudia!"

Claudia Simpson, age eleven, strolled into the bathroom and sniffed. "Yes," she said, cryptically. "I suspected as much. Mr. Blastwood died by his own hand."

"Suicide?" Yoghurty squeaked.

"No. Stupidity. Surely you know about chemical compounds."

Yoghurty scratched his head. "I must have been absent that day," he said.

"Have you ever put raisins in your cereal?" asked Claudia. "Of course you have, and you know how easy it is to remove them. In fact, if you want, you can pick out all the raisins and eat them first. That's because cereal with raisins is a **mixture**. You create a mixture when you combine two or more substances that can then be easily separated. In the raisin mixture, both the raisins and the cereal don't lose their individual **properties**, or characteristics, just because you combine them."

Yoghurty blinked dumbly. "Are you saying that Mr. Blastwood was killed by raisins?"

Claudia clicked her tongue. "Don't be dense," she said and turned on the faucet in the sink. "Do you see this? It's water."

"Yes, thank you," said Yoghurty, "but I'm not thirsty."

"Water is made of the **elements** hydrogen and oxygen, which are both gases, but notice that water, a liquid, looks and acts nothing like the two gases of which it is made. In fact, hydrogen is explosive, and oxygen helps things burn. But when combined, they can put out fires."

"Cool!" said Deputy Sikes.

"Cool, indeed," said Claudia. "So water is not a mixture, but a **compound**. In a compound, the elements combine in such a way that they change their properties and create something totally different from the original elements. For instance, sometimes two or more gases combine to create a liquid."

"Like water!" cried Deputy Sikes, eager to prove that he had been listening.

"And sometimes," said Claudia, "the resulting compound is even stranger than that. Sodium is a metal, and chlorine is a poisonous gas, but when you put them together as a compound you get ordinary, perfectly harmless table salt!"

"Yes, yes, yes," said Yoghurty. "That's all very interesting, but it doesn't explain why Blastwood, here, is as dead as a mackerel."

Claudia pointed to a white plastic bottle. "There's one clue," she said. "Apparently, Mr. Blastwood was using chlorine bleach to clean his bathtub because one of the properties of bleach is that it whitens things. People add it to their wash all the time. But Mr. Blastwood was not content to simply use the bleach. He must have decided he wanted what he thought would be a more powerful cleaner."

"And with good reason," said Deputy Sikes. "The glass doors enclosing the tub are really gooey."

"Exactly," said Claudia. "So Mr. Blastwood got the idea to mix ammonia with the bleach, but by doing this, he created a deadly chemical reaction. The chlorine that was safely combined with the sodium, hydrogen, and oxygen in the bleach was  suddenly released as chlorine gas! Chlorine gas is extremely, extremely poisonous, and it killed him." She smacked the sink. "Blam! Just like that."

"Gosh!" breathed Deputy Sikes.

"Well, I for one am going to stay out of bathrooms," said Yoghurty, standing up and rubbing his aching knees.

"And I'm never going to clean mine," promised Deputy Sikes.

Claudia shook her head. "No, what we need to do is warn people not to go messing around with the chemicals in their homes. Some, as you see, are deadly when combined. In fact, it says so right here on the label of the bleach bottle. Mr. Blastwood should have taken the time to read it, or at least paid more attention in his science class."

"Coming through!" shouted the medical examiner.

Yoghurty, Sikes, and Claudia filed out of the bathroom so the late Mr. Blastwood could be prepped for his final ride downtown.

"She's pretty good," said Sikes as Claudia strolled off into the setting afternoon sun.

"Yeah," said Yoghurty. "I taught her everything she knows."

Name _____

Questions for
A Compound Mystery

Answer the following questions in complete sentences.

1. Why is a can of peas and carrots a mixture?

   _____

   _____

2. What other mixtures can you think of? List four more.

   _____   _____

   _____   _____

3. The composition of several common compounds is mentioned in the story. Match each common compound with the elements that form it. Then circle the element that killed Osgood Blastwood.

   water   sodium + hydrogen + chlorine + oxygen

   salt   hydrogen + oxygen

   bleach   sodium + chloride

4. Explain how Osgood Blastwood was killed.

   _____

   _____

   _____

   _____

Claudia's Word Mixture

Each of these phrases describes a word that plays an important role in the story. Identify the word and then circle it below. Words run left to right, right to left, and diagonally.

1. two or more substances that, when combined,
   can be easily separated   _____

2. an explosive gas   _____

3. sodium + chlorine   _____

4. a poisonous gas   _____

5. what you should never mix with bleach   _____

6. the characteristics of an element   _____

7. a gas that promotes burning   _____

8. a substance made from two or more substances   _____
   that have combined chemically

```
A D X T R V B N I T R O G E N L M O Q
H Y D R O G E N S Q W T R E C B C V L
D V B A R T Y U V X N E G Y X O B M A
N T T U M I X T U R E A X N O L G E N
U A L S A M U L N B M V T B E W Q T U
O E C H U P O E N I R O L H C F L C X
P W Q A Z C B N L O P I A S D F G H J
M Z C B C L A E I I A C S S U L F U R
O S C Y C B T D E A T Z T A L S V O X
C N K K Z S E I T R E P O R P S A L I
```

Note: This is an excellent activity for learning groups.

Kitchen Chemistry

What's your specialty? A cheese omelet with mushrooms and peppers? Five-alarm chili? Chocolate pudding with nuts and raisins?

The American Society of Chemists would love to hear about your favorite recipe. Unfortunately, they're all terrible cooks and will only understand it if you describe it in terms of mixtures and compounds.

1. List each ingredient and tell why you think it is a mixture or a compound.

2. Explain how the ingredients are combined and if the result is a mixture or a compound.

3. What is the final result—a mixture or a compound?

4. Use diagrams when necessary.

A Breath of Wind

A tornado was in the works. Ethan could smell it. Something always happened to the air. It got heavy and as still as death, and the sky turned a funny shade of green. Far off, Ethan could hear the growl of thunder. A huge flat-topped **cumulonimbus** cloud—a thundercloud, a cloud that makes tornadoes—hung over Jake Dawson's fields like a UFO.

The screen door creaked on hinges that Ethan kept forgetting to oil. "Lemonade, Ethan?"

Ethan nodded, and his wife returned to the kitchen.

"We'll be needing the storm cellar, I believe!" he shouted in to her.

The tornadoes always came to Oklahoma in the early spring. A southern sun heated the waters in the Gulf of Mexico, and the warm, moist air rose and pushed northwest over Texas and on into Oklahoma and Kansas. Cool air dropped down from Canada. Dry, upper-level air flowed east over the Rockies and spilled onto the plains. Then the air masses began their dance. Slowly, they drew closer together, finally colliding at an invisible line called a **front**.

Like great whales, the air masses would slide past each other. As the faster, upper-level winds moved over the slower winds below, the air between them would begin to roll. All along the front, **thunderheads** would start to take shape. The clouds would churn and mix, flowing and twisting and working themselves into full-blown thunderstorms. Sometimes there'd be a whole **squall line** of thunderstorms that might stretch as far as a hundred miles. Ethan had lived in Tornado Alley long enough to know that one of those thunderstorms could easily grow into a monster. The weather folks called it a **supercell**, but that was just a fancy word for a cloud filled with nothing but trouble.

Ethan heard the screen door slam and the clink of ice as Sarah set down the pitcher of lemonade. "They posted a watch," she said.

A tornado watch. Ethan nodded. He'd figured as much. "Sky's getting that funny color," he said. He squinted at the eerie green and swallowed lemonade.

The wind was picking up. Sarah's chestnut hair had begun to swirl about her, and the long grasses in the front yard rippled like waves on the sea.

The winds inside the supercell that sat above Jake Dawson's fields were spinning now. Most of these little **mesocyclones** would die out, unraveling like a curl that goes limp in hot weather. But today one of them would keep on spinning, drawing up more and more warm, humid air from the ground, faster and faster....

"There!" shouted Ethan.

A half-formed tube had dropped from the churning thundercloud. A second later it broke apart, reformed, dissolved again, and finally found its deadly funnel shape. It had made contact with the ground and was spinning counterclockwise, inhaling everything as it thundered across the land. Horrified but fascinated, Ethan and Sarah watched the tornado demolish Edie Briggs' greenhouse. Hundreds of geraniums were sucked in, and for a few eerie moments, the funnel blushed geranium pink.

Sarah's blue cane chair began to rock, and the wind chimes that hung from the porch roof jangled in panic.

Sarah had thrown open the storm cellar door. "Ethan!" she screamed. Her hair was wild now, whipping around her face.

Ethan was frozen, unable to take his eyes off the tornado as it rushed forward, tossing Hattie's red pickup truck into the air, swallowing the barn Waylon had finished building just last week.

"Ethan!" Sarah screamed again, as a bird feeder was torn from a tree. "Don't make me stay out here and die with you!"

And then he was running, pushing Sarah into the safety of the storm cellar and slamming the wooden doors behind him. The roar was deafening, like a train carrying freight to Oklahoma City. They huddled in the darkness, their eyes shut, holding each other, praying for the tornado to spare the house, to spare the animals, to spare their lives. The tornado pulled at the storm cellar doors, wrenching them free and carrying them away as it bellowed like a T-Rex, and then…

Just like that, it was gone.

Shaking, they climbed the cellar steps. The giant, old hickory tree was in splinters, and the fence was missing. The front yard was littered with all kinds of broken junk. Some of it looked like Jake's. Half of the roof had been ripped away, along with the two front bedrooms. But Ethan could hear the cows lowing in the pasture, and it was like music. He took Sarah's hand and pointed to the pitcher that sat, impossibly unbroken and untouched, on the front porch steps.

"How lucky we are," he whispered, as he kissed her cheek. "We can finish our lemonade."

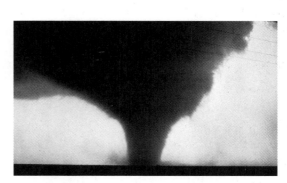

Photo: National Oceanic and Atmospheric Administration/Department of Commerce

The tornado funnel sucks up everything in its path as it moves across the land.

Questions about
A Breath of Wind

Fill in the bubble that best answers each question.

1. What sort of clouds are associated with tornadoes?

   ○ cirrus
   ○ stratus
   ○ cumulonimbus
   ○ altocumulus

2. What are mesocyclones?

   ○ large, dangerous tornadoes
   ○ funnel-shaped tornadoes
   ○ severe thunderstorms
   ○ spinning winds that form inside supercells

3. Where would you expect to have the greatest chance of experiencing a tornado?

   ○ in Florida
   ○ in Kansas
   ○ in New York
   ○ in California

4. What made the tornado's funnel appear pink to Ethan and Sarah?

   ○ light from a red setting sun filtering through the cloud
   ○ sunlight reflected off water droplets in the funnel
   ○ the red geraniums it had sucked up
   ○ dust swirling in the center of the funnel

5. Why do most tornadoes in the U.S. occur in Tornado Alley?

   ○ The land is narrow there, so the wind goes faster.
   ○ The weather is warm there in the summer.
   ○ Mixing air masses create the right conditions.
   ○ Tornadoes only form where the land is flat.

Vocabulary

A. What Did You Mean by That?

Some words have more than one meaning. The meaning depends on how the word is used in the sentence. What do the underlined words mean in these sentences from the story? Circle your answer.

1. The sky turned a funny <u>shade</u> of green.

   a covering for a window   a color

2. All along the <u>front</u>, thunderheads would start to take shape.

   the place where air masses meet   the forward part

3. They posted a <u>watch</u>.

   a notice or bulletin   a timepiece

4. But today one of them would keep on spinning, <u>drawing</u> up more and more warm, humid air.

   creating a picture   pulling upward

B. Bringing the Tornado to Life

The author says the tornado *swallowed* a barn. Swallowing, of course, is something only living creatures can do, but describing the tornado in this way really brings it to life. What other words and phrases does the author use to give the tornado the characteristics of a living being? List as many as you can find.

An example has been done for you.

The tornado rushed forward, swallowing the barn.

_____

_____

_____

_____

Set the Stage for a Tornado to Form

Reread the information in the story about how tornadoes form. Study the map below.
Then number the events in order.

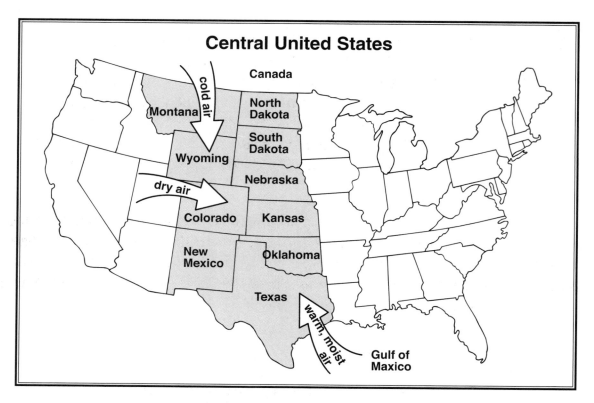

Central United States

Canada

Montana

cold air

North Dakota

South Dakota

Wyoming

Nebraska

dry air

Colorado

Kansas

New Mexico

Oklahoma

Texas

warm, moist air

Gulf of Maxico

_____ As the air masses collide, the air churns and mixes, forming clouds called thunderheads.

_____ Warm, moist air from the Gulf of Mexico collides with dry air from the Rocky Mountains and cold, polar air from Canada.

_____ As the storm moves across the land, the funnel moves with it, sucking things up into itself.

_____ If conditions within a thunderhead are right, the moving air masses begin to spin rapidly.

_____ The spinning air spreads out, making a funnel of air, with the small part of the funnel reaching toward the ground.

An Inside Look at the San Andreas Fault

If you've ever lived in or visited California, you've probably heard of the San Andreas Fault. Earthquakes along the San Andreas Fault have created some of California's best scenery and worst trouble. In order to understand how these "big shakes" happen, you need to know a bit about what lies far beneath your feet.

Earth's crust is made up of solid sections of rock called tectonic plates that float and slide on Earth's molten mantle. Sometimes one plate's edge slides under another plate. Deep trenches on the ocean floors are proof of this kind of movement. Sometimes two plates collide, making mountains. Sometimes plates slide past each other. Movements of tectonic plates cause faults, or large breaks, in the crust.

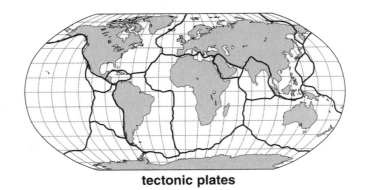

tectonic plates

The San Andreas Fault is in western California. It is more than 650 miles (1,046 km) long and 10 miles (16 km) deep. It extends from north of San Francisco southward to San Bernardino. It is the boundary of two of the Earth's moving plates, the Pacific plate on the west and the North American plate on the east.

These two plates creep at the slow rate of a few centimeters a year. They have moved only 350 miles (563 km) in the past 20 million years. As they move, they slide against each other. At some places along the fault, this slide is slow and continuous. This even, steady creep does not cause earthquakes. At other points along the fault, the rocks of the plates get caught on one another as they slide. For one hundred or more years at a time, these "locked" sections do not move at all. Over time, pressure builds up in these areas.

Then the strain is released in a single lurch. When this happens, Earth's crust snaps into a new position. This sudden "faulting" causes vibrations that are felt as earthquakes. The first vibration waves produce a "thud." The next set of waves make the ground roll and sway.

The ridges and valleys of the San Andreas Fault can be seen easily from the air. From the ground its features are less striking. People travel, live, and do business within the fault zone without knowing it. Yet, if they look closely at the landscape, they can tell they are in the zone. Streams make sudden right turns when they cross the fault line. In some spots along the fault, the vegetation and terrain look different on one side of the fault than on the other. High, narrow ridges surrounding deep, still ponds are another sign of the fault zone. In some places along the fault, observers can even see offset fences, roads, and rows of trees moved by earlier earthquakes.

The San Andreas Fault was discovered in 1893 when geologist Andrew Lawson took a close look at the landscape. He found signs of earth movement all along the way from San Diego to San Francisco. Lawson defined the borders of a fault. He named it the San Andreas because its features were most clear around San Andreas Lake.

Thousands of tiny earthquakes occur along the San Andreas Fault each year. Two of the strongest earthquakes in recent history occurred in 1857 and 1906. The 1857 earthquake struck Southern California. No towns were located near the center of the quake so little damage was done to buildings or people.

The 1906 earthquake caused more damage. It occurred in San Francisco where many people lived and worked. The shaking of the quake knocked down buildings. It also broke power lines and overturned wood stoves, causing fires. The fires spread quickly through the wooden structures of the city. More than 700 people died in the disaster. Thousands more were left homeless. Much of San Francisco had to be rebuilt from scratch.

Today we know how to construct buildings that are less likely to fall or burn in earthquakes. We know which kinds of soil are safe to build on. We even have instruments that help us predict when and where earthquakes might occur. Living in the San Andreas Fault zone is much safer today than it was in 1906.

Questions about
An Inside Look at the San Andreas Fault

Fill in the bubble that best answers each question.

1. Which of these is **not** a land feature of the San Andreas Fault?

   ○ sudden turns in streams
   ○ deep, still ponds
   ○ numerous purple wildflowers
   ○ high, narrow cliffs

2. Why are some sections of the San Andreas Fault called "locked"?

   ○ Geologists cannot do research on locked sections because they are in deep wilderness.
   ○ Tourists cannot visit locked sections because no roads lead to them.
   ○ Locked sections of the San Andreas Fault have fences around them.
   ○ Locked sections of the fault do not move for a hundred or more years at a time.

3. Which movement listed is **not** a movement that tectonic plates experience?

   ○ Plates collide.
   ○ Plate edges slide under one another.
   ○ Plates slide past each other.
   ○ Plates creep at the rate of one mile per year.

4. What causes earthquakes?

   ○ the steady, even creeping of tectonic plates sliding past each other
   ○ locked sections of a fault moving suddenly and sending out vibrations
   ○ heavy rock avalanches that send out vibrations
   ○ unusually strong wave action in the world's major oceans

5. How many earthquakes occur in California each year?

   ○ about five
   ○ no more than one hundred
   ○ thousands of major earthquakes
   ○ thousands of tiny earthquakes

6. Why was the 1906 earthquake in Northern California more destructive than the 1857 earthquake in Southern California?

   ○ The 1906 earthquake burst water lines causing a major flood to follow the earthquake.
   ○ The Northern California earthquake occurred in an unpopulated area.
   ○ The Southern California earthquake occurred in an unpopulated area.
   ○ The 1857 earthquake did not last as long.

Vocabulary

Find words in the story to complete the puzzle.

Across

2. a vibration caused by sudden movement along a fault
3. the outer layer of the Earth
5. something that causes great trouble
6. a large break in the Earth's crust
7. a tectonic plate west of the San Andreas Fault
8. sections of the Earth's crust
9. a major earthquake fault in California
10. a scientist who studies the formation of the Earth

Down

1. plant life
4. a California city struck by a major earthquake in 1906

Word Box

crust	Pacific plate
disaster	San Andreas
earthquake	San Francisco
fault	tectonic plates
geologist	vegetation

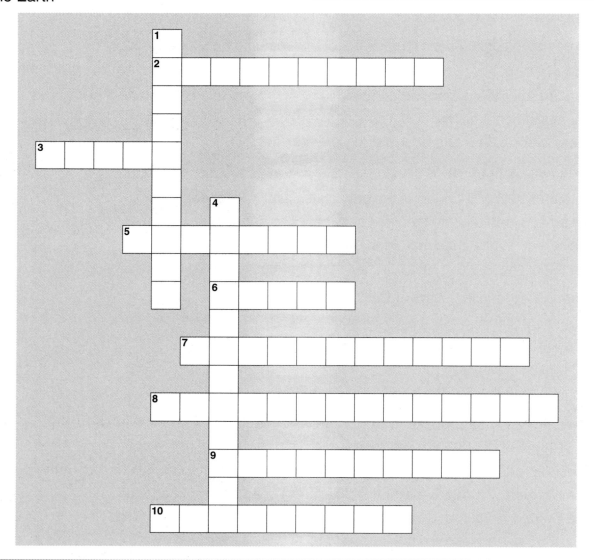

Locating Tectonic Plates

Geologists look at where earthquakes, volcanoes, and mountain ranges occur to find out where tectonic plate boundaries might be. Look at the map of the major tectonic plates below. Then circle **T** or **F** below to show which of the statements about the map are true. You may need to use an atlas or a globe to help you locate specific places.

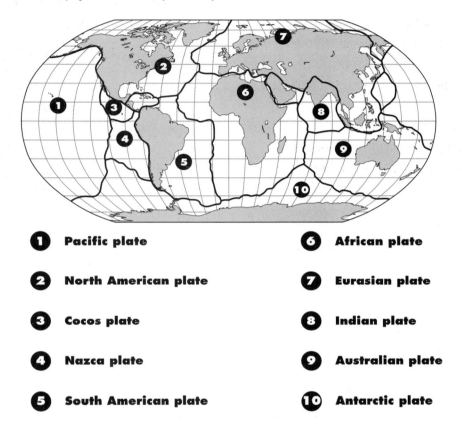

❶	Pacific plate	❻	African plate
❷	North American plate	❼	Eurasian plate
❸	Cocos plate	❽	Indian plate
❹	Nazca plate	❾	Australian plate
❺	South American plate	❿	Antarctic plate

T   F   1. The Pacific plate does not include a continent.

T   F   2. The South American plate includes all of South America.

T   F   3. Most of the United States is included in the North American plate.

T   F   4. China is located on the Indian plate.

T   F   5. The African plate borders the Pacific plate.

T   F   6. Most plate names are related to their locations.

Dear Professor Parsec

Dear Professor Parsec,

I love the stars and am interested in traveling to a few of them, but I can't seem to get off the ground. How can I get my rocket into space and keep it there?

Cosmically yours,
Star Commander Lance Redshift

Dear Lance,

I'm asked this question all the time. In principle, getting into space is easy. All you have to do is get your rocket moving fast enough to reach escape velocity, and off you go.

Ah, but what's escape velocity?

Everything on Earth, as I'm sure you know, is held down by gravity. You toss a ball into the air, and what does it do? It goes only as high as the force of your toss will allow and then falls back down to Earth. That's gravity for you, always showing off how attractive it is. But Earth's gravity can be overcome if you push your rocket's speedometer up to 7 miles per second—the velocity needed to escape our planet. Doing a little math, here...that's equal to 420 miles per minute, or a mere 25,200 miles per hour. Hit that speed and wave good-bye to Earth because you're on your way to the stars.

There is, however, one detail that you really must consider: the weight of your rocket. Have you ever tried to push a freight train? Well, no, maybe you haven't, but if you're ever tempted, bear in mind that the heavier something is, the tougher it's going to be to get it moving. (That certainly makes a lot of sense, doesn't it?) In fact, the great astronomer Sir Isaac Newton figured this out long ago and made it part of his three Laws of Motion. He said, "An object at rest tends to remain at rest." He described this unwillingness to move as inertia. So there's your rocket on the launch pad, and it's suffering from a bad case of inertia. How do you achieve lift-off?

Thrust, my boy! Thrust! Thrust is the oomph, the force you need to get your couch potato rocket moving. This is where the rocket engines and the fuel enter the picture.

Now, fire needs oxygen. Since there's no air in space, nothing can burn. So you can't use, say, a suped-up car engine in your rocket because car engines operate by burning gasoline. Ah, but the rocket scientists have come up with an engine that runs on a combination of liquid oxygen—LOX for short—and liquid hydrogen. This way they can bring the oxygen with them. The chemicals are loaded into enormous fuel tanks and attached to the rocket in what are called stages.

Now we move to the launch pad and start the engines. As the engines run, they use fuel and, just like a car, they create exhaust, which shoots out the rear end of the rocket. It's the force of the exhaust pouring out of the tail end of the rocket that creates the thrust to lift the rocket. And here comes Newton again with yet another Law of Motion. This one promises that if you send the exhaust out in one direction with a certain amount of force, the rocket will move in the opposite direction with the same amount of force. So if the exhaust is moving in a downward direction, the rocket is moving UP!

Ready? Fasten your seat belts. Here we go!

The engines are screaming, and the rocket is beginning to lift. More and more fuel is being used to get the rocket up to escape velocity: 5½ miles per second...6 miles per second...6½... 7 miles per second...and you're free of Earth's gravity!

And now, thanks again to Sir Isaac Newton, you are truly star bound. You can coast to the stars and never have to use another drop of fuel because Newton's very first Law says, "An object in motion remains in motion unless something acts upon it to slow it down or speed it up." Out here, in the icy silence of space, there is nothing, not even air, to affect the speed of your starship. You can cruise forever, moving easily at 7 miles per second or, if you're in a hurry, you can burn more fuel and go faster.

So reach for the stars, Lance, and have a great trip.

Your pal in space,
Professor Parsec

Rocket

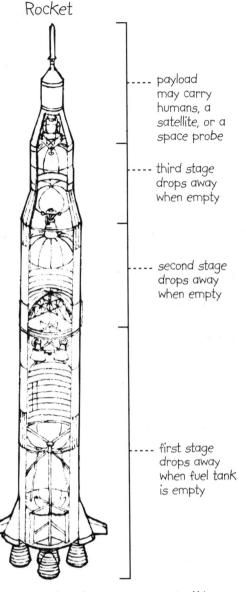

- - - payload may carry humans, a satellite, or a space probe

- - - third stage drops away when empty

- - - second stage drops away when empty

- - - first stage drops away when fuel tank is empty

Dear
Professor
Parsec

Questions for
Dear Professor Parsec

Fill in the bubble that best answers each question.

1. Which of the following helps a rocket to overcome inertia?
   - ○ thrust
   - ○ reaching escape velocity
   - ○ liquid oxygen
   - ○ less air

2. Lance Redshift drew four different designs for his rocket. Which one should Lance build if he wants to get his rocket off the ground?

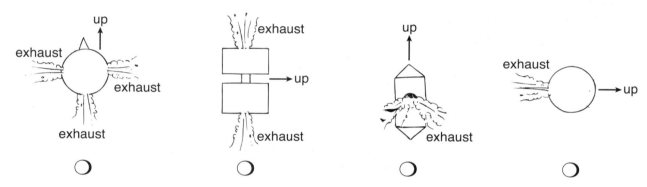

   ○   ○   ○   ○

3. Why does Lance have to carry liquid oxygen aboard his rocket?
   - ○ It provides a source of air for the passenger compartment.
   - ○ It supplies oxygen so the fuel can burn.
   - ○ It allows Lance to achieve escape velocity.
   - ○ It prevents the rocket engines from overheating.

4. The Moon's gravity is not as strong as Earth's. What does that mean if Lance wants to launch his rocket from the Moon?
   - ○ His escape velocity will have to be greater than 7 miles per second.
   - ○ His escape velocity will be less than 7 miles per second.
   - ○ His escape velocity will be the same as Earth's.
   - ○ He will not be able to launch his rocket from the Moon at all.

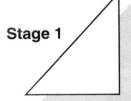

Vocabulary

Word Launch

Write the word or words being defined in the squares to build the rocket from the ground up.

Stage 1

    a. Newton's _____ of Motion

    b. 7 miles per second (from Earth)

Stage 2

    c. the force of this creates thrust

    d. an object at rest tends to remain at rest

    e. apples fall down because of this

Stage 3

    f. this helps a rocket overcome inertia

    g. LOX stands for this

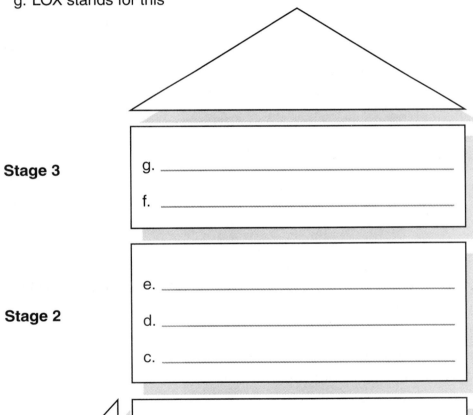

Stage 3

g. _____

f. _____

Stage 2

e. _____

d. _____

c. _____

Stage 1

b. _____

a. _____

          Read and Understand, Science • Grades 4–6 • EMC 3305

Dear Professor Parsec

Professor Parsec received these three letters while he was away on vacation. Fill in for the professor by answering the readers' questions. Don't worry. Everything you need to know is in the professor's response to Lance.

Dear Professor Parsec,

    I built my rocket on the front lawn and want to move it to the driveway, but I can't budge it. Why not? Is there some law against it?

_____

_____

_____

Dear Professor Parsec,

    Help! I started the engines in my rocket, and they seemed to work okay, but my rocket only rose about ten feet and then crashed. What went wrong?

_____

_____

_____

Dear Professor Parsec,

    I hope this e-mail arrives in time. I had to use all my fuel to get my rocket up to 30,000 miles per hour. How can I prevent it from slowing down in space?

_____

_____

_____

Jocelyn Bell Burnell

Jocelyn Bell Burnell

Jocelyn Bell Burnell likes a challenge. When the Irish-born Burnell was a student at Cambridge University in England, her professor said they needed a **radio telescope**. So, she and a handful of other students built it. The project took two years to finish. When it was done, it would scan the sky for **radio signals** from the stars.

The radio telescope was huge. It covered four and a half acres (1.82 ha). To build it, they strung wires and **antennae** between tall poles. In all, there were more than a thousand poles. The poles held more than 120 miles (193 km) of wire and 2,000 antennae.

Before it was even quite finished, Jocelyn began work. The radio telescope constantly printed out long sheets of chart paper. Radio signals from space showed up as marks on the paper. Jocelyn looked at every mark. Every complete scan of the sky took about four days and about 142 yards (130 m) of chart paper. That's more than enough paper to reach from one end of a football field to the other!

One day, she saw something odd. It was a mark about half a centimeter long. It didn't look like something that would come from a star. It didn't look like a radio signal from Earth, either. At first she thought it might be **scruff**. Scruff is radio **interference**. It's sort of like the white lines of static that sometimes show up on television sets.

Jocelyn watched to see if the mark showed up again. At first she saw nothing. Then, after several weeks, it was there again. It was from the same part of the sky as before. After looking at the mark closely, Jocelyn knew it was not scruff. It seemed to show several pulses of radio signals. The pulses came every 1.3 seconds. This was too regular to be scruff.

In the months that followed, Jocelyn and her professors tried to figure out what the signals meant. Could they be signals from Earth, bouncing around in space? No. They seemed to come from outside our solar system. Could they be made by beings on another planet? Jocelyn jokingly called them "LGM." It stood for "little green men."

As it turned out, they weren't messages from aliens. They were from a new kind of star. Jocelyn had made a major discovery. She had found the first **pulsar**.

When a very large star explodes and dies, the center of the star collapses. It becomes very dense and is called a **neutron star**. Most scientists think that a pulsar is a neutron star that spins very fast. With each spin, it sends out a flash of radio waves. Think of a lighthouse. On this lighthouse, the lantern turns around and around. With each turn, it throws a flash of light at you. That's kind of like what a pulsar does. Instead of light, it throws radio waves. The marks that Jocelyn found on her chart paper were from these flashes of radio waves.

The world's scientists buzzed with excitement. In 1974 Jocelyn's professors received the **Nobel Prize** for their part in the discovery. (Some people think Jocelyn should have shared the prize. She feels that things were done in a fair way.) That first pulsar is now called CP1919. She and other scientists have gone on to find about a thousand more.

At 24 years old, Jocelyn had found a new kind of star. What a way to begin a career in astronomy! For many years afterward, she chose to work part-time. She made this choice in order to have more time for her family.

Even part-time, she managed to get a lot done. She worked with **gamma-ray** and **X-ray telescopes**. She watched the stars with telescopes fixed on special balloons. She put telescopes on **satellites** and rockets. She used telescopes set high in the mountains of Hawaii. Often called on to make speeches, Jocelyn has won many awards. She also teaches classes at England's Open University.

Now, after more than 30 years of scientific discovery, Jocelyn is still excited about it. In her words, "**Astronomy** is terrific!"

What Is a Pulsar?

A pulsar is a collapsed star that throws out radio waves as it turns.

Pulsars are good timekeepers. Although each one spins at a slightly different speed, most turn about two times per second. Each turn takes exactly the same amount of time as the one before it.

Name _____

Questions about
Jocelyn Bell Burnell

Fill in the bubble that best completes each sentence or answers each question.

1. To build the radio telescope, Jocelyn and the other students _____.

   ○ connected a radio and a telescope together
   ○ placed a large telescope on top of a building
   ○ strung wires and antennae between tall poles
   ○ wired mirrors to the tops of tall poles

2. What happened when the radio telescope picked up radio signals from space?

   ○ The signals showed up as marks on chart paper.
   ○ Beeping sounds came from the radio.
   ○ Jocelyn took a picture of that part of the sky.
   ○ A computer assigned each signal a number.

3. What did Jocelyn do immediately after finding the unusual signal?

   ○ She told a reporter it must be "little green men."
   ○ She waited to see if it would happen again.
   ○ She worked with gamma-ray and X-ray telescopes.
   ○ She won the Nobel Prize for her discovery.

4. What was Jocelyn Bell Burnell's important discovery?

   ○ a type of radio telescope
   ○ a new kind of star called a pulsar
   ○ radio interference from space
   ○ a star that was collapsing

5. According to the article, pulsars _____.

   ○ reflect little light
   ○ are easy to see
   ○ don't last very long
   ○ are good timekeepers

Name _____

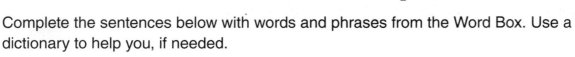

Vocabulary

Complete the sentences below with words and phrases from the Word Box. Use a dictionary to help you, if needed.

Word Box				
solar system	pulses	regular	dense	antenna
pulsar	radio telescopes	astronomy	collapse	antennae

1. If you are interested in stars and events that happen in space, then you should study the

   field of _____.

2. Our _____ includes the sun and all the objects that orbit it, including the nine planets.

3. A _____ is a fast-spinning type of neutron star discovered by Dr. Jocelyn Bell Burnell.

4. Most pulsars send out flashes, or _____, of radio signals about two times per second.

5. Astronomers use _____ to scan the sky for radio signals from space.

6. Neutron stars are very compact, or _____.

7. To receive radio or television signals, it is helpful to have an _____.

   The word for more than one of these is _____.

8. When something falls in on itself, it is said to _____.

9. Something that follows a pattern, happening again and again with the same amount of time

   between each occurrence, is said to be _____.

Jocelyn Bell Burnell

1. From a very young age, Jocelyn enjoyed learning about the stars. What do you enjoy learning about?

_____

_____

_____

2. When Jocelyn discovered the unusual radio signal, she was puzzled and curious. Her curiosity led her to a big discovery. Write a story, true or make-believe, about a time when your curiosity led you to learn about something.

_____

_____

_____

_____

_____

_____

_____

_____

_____

_____

_____

_____

Probing the Periodic Table

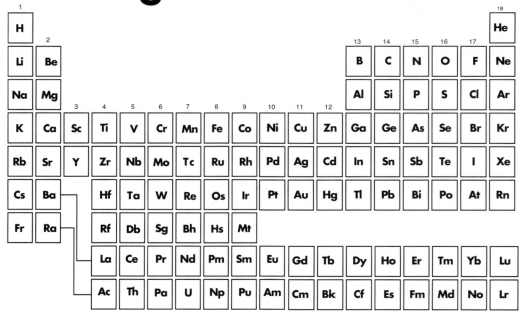

Periodic Table of Elements

Elements are the basic building blocks of all things. Water is a combination of the elements hydrogen and oxygen. Silicon is an element used in computer chips. Gold and silver are elements used to make jewelry. The element carbon is found in all living things.

Elements are made of **atoms**. Atoms are tiny. You need a special microscope to see atoms. Even though they are small, atoms have weight. The **atomic mass** of an element is the weight of one atom of the element.

In 1869 Dmitry Mendeleev listed all the elements known at the time in order of their atomic weight. He arranged his list into a table of columns and rows. He started his table with the lightest element, hydrogen. He continued to add elements, moving from lightest to heaviest.

When he had finished, Mendeleev noticed patterns in his chart. The traits in each row of elements changed gradually. That is, the second element on his chart was very similar to the first. The third one was slightly less similar, and so on. Mendeleev noticed other patterns in his chart. At regular intervals, traits of elements were repeated. For example, every fifth element might be shiny. Every eighth element might be liquid at room temperature.

Mendeleev also noticed gaps in his chart. Sometimes an expected trait was missing at a regular interval. Mendeleev guessed elements would one day be discovered to fill his gaps. He left blanks in his chart where he expected new elements to fit. Soon new elements were found. They fit the blanks left in Mendeleev's table.

In Mendeleev's day there were 63 known elements. Today we know of almost 100 elements that can be found in nature. These elements are listed in a chart very much like Mendeleev's. The chart is called the **periodic table**. On the modern chart, elements are listed by **atomic number** instead of atomic mass. We know about atomic numbers today because we have learned about the parts that make up an atom. All atoms are made up of three types of smaller particles: **protons**, **electrons**, and **neutrons**. Each element has a specific number of protons in the nucleus of its atom. An element's atomic number is the number of protons inside one atom of the element. When elements are listed in order of their atomic number, the patterns Mendeleev discovered are even more clear.

Each square on the periodic table tells about one element. The number at the top of the square is the atomic number. The number at the bottom of the square is the element's atomic mass (weight). Above the atomic mass, the element's name is listed. The big letter, or pair of letters, in the middle of the square is the element's symbol.

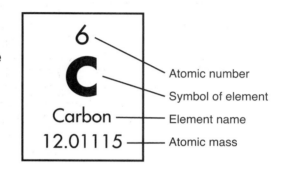

The seven horizontal rows of the periodic table are called **periods**. The number of protons and electrons in the atoms of elements in each period increases from left to right. The number of electrons in the atoms of the element decides an element's traits. That's why the traits of elements change gradually in each row.

Vertical columns in the periodic table are called families or **groups**. Groups are numbered from left to right. Elements in groups are closely related. For example, most elements in Group 1 are shiny and soft. They explode when mixed with water. They are known as alkali metals. Members of Group 18 are gases that do not like to mix with other elements. They are known as the noble gases.

Large sections of the periodic table share traits too. The elements on the left side of the table are all metals. Metals share many traits. They are shiny. They can be hammered or stretched without breaking. Electricity and heat travel easily through them. Nonmetals on the right side of the chart do not have these traits. They are not shiny. They do not change shape easily. They do not conduct electricity or heat well.

Chemists and students use the periodic table as a reference tool. The chart distinguishes metals from nonmetals. It tells the mass and symbols of elements. It shows which elements share traits. For these reasons, it is a useful tool in the laboratory.

Periodic Table

Questions about
Probing the Periodic Table

Answer the following questions in complete sentences.

1. What are elements?

    _____

    _____

2. What is the periodic table?

    _____

    _____

3. Who published the first version of the periodic table?

    _____

    _____

4. What information is included on each square of the periodic table?

    _____

    _____

    _____

5. What is a group or family on the periodic table?

    _____

    _____

7. How do chemists use the periodic table?

    _____

    _____

    _____

    _____

Vocabulary

A. Words from the article are written below with missing letters. Read the clues to help you fill in the missing letters.

1. groups are made of these columns that go up and down

    vert ____ ____ ____ ____
                  x

2. these parts of an atom decide an element's traits

    elec ____ ____ ____ ____ ____
         x

3. Mendeleev's chart ordered elements by this

    ato ____ ____ ____ w ____ ight
                   x

4. this particle is even smaller than an atom

    pro ____ ____ ____
           x

5. horizontal rows on the periodic table chart

    ____ ____ ____ iod ____
               x

6. a chart that lists elements

    ____ ____ ____ iodic ta ____ ____ ____
         x

7. the letter or pair of letters that stand for an element in a periodic table square

    sy ____ ____ ____ ____
       x

8. the number at the top of each square in the periodic table

    atom ____ ____ numb ____ ____
                   x

B. Now write the letters that have an **x** under them:

____ ____ ____ ____ ____ ____ ____ ____

Unscramble the letters to complete this sentence:

The periodic table is an arrangement of the _____ found in nature.

Interpreting the Table

Each square on the periodic table tells about one element. Look back at the article to remind yourself what each number and letter stands for. Then answer the questions about the squares in the portion of the periodic table shown here.

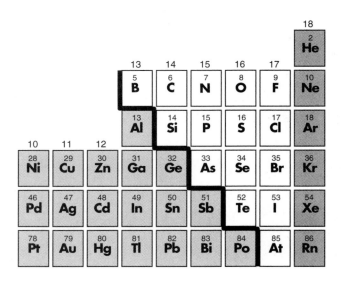

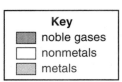

1. Including the noble gases, how many nonmetals are listed on this section of the periodic table?

   _____

2. Neon is a noble gas. What is the symbol for Neon? (Hint: The symbol for an element is usually related to its spelling.)

   _____

3. Elements touching the zigzag line show properties of both metals and nonmetals. They are called metalloids. Give the symbol for three metalloids.

   _____

4. Gold is element 79. What is its symbol?

   _____

5. What element shown has the highest atomic number?

   _____

6. What elements are in family 17? Write their symbols.

   _____

The Mystery of the Melted Candy Bar

**Percy Spencer
1894–1970
Inventor of the
microwave oven**

Friends of the young American Percy Spencer might not have guessed the boy would go far in life. After all, he was orphaned as a child and only finished the sixth grade. Still, Spencer was smart and curious, and those traits landed him a job in a science lab. Once there, Spencer set out to make up for his lack of schooling. He questioned and tested. He listened and learned. Before long, the self-taught Spencer was a top engineer. By 1941 he had patented a hundred inventions.

That's why the British came to Spencer during World War II. America and Britain were fighting on the same side in that war. Britain had devised a tube called a **magnetron** that produced **microwaves**. The microwaves bounced off the metal of ships and returned a signal. So the magnetron acted as radar. It could find Nazi warships in the dark. But it was hard to build magnetrons. Only a few could be made each day—until Britain turned to Spencer. Spencer showed the British how to mass-produce the magnetron. Now more than two thousand could be made per day. Spencer was cheered as a war hero and genius.

After the war, Spencer worked to improve the radar system. That's all he planned to do. He was not looking for a new way to cook food quickly. But then a strange thing happened. One day when Spencer was working around a magnetron tube, he noticed that a candy bar in his shirt pocket had started to melt even though he felt no heat. Spencer was curious. He placed popcorn kernels beside the magnetron. The **microwaves** that escaped the tube popped the corn. He held an egg close to the magnetron. The shell burst, splattering cooked egg yolk in Spencer's face.

Spencer's egg cooked because water in the egg absorbed microwaves. The energy in the microwaves made the water molecules in the egg begin to vibrate. The vibration of the water molecules caused the temperature of the egg to increase. The increase in temperature cooked the egg.

Spencer thought microwaves could be useful in the kitchen. He built a metal box in which he could place food. He channeled microwaves into the box. The microwaves cooked the food. After two years of testing, Spencer's microwave oven hit the market. It was called the **Radarange**.

 Read and Understand, Science • Grades 4–6 • EMC 3305

The Radarange worked, but it was not perfect. The first models were six feet tall and heavy. The magnetron inside the Radarange was cooled with water, so plumbing was needed to make it work. Each oven cost $3,000.

Only large restaurants, railroad cars, and ships used the first Radaranges to cook food. They were also used in factories. The huge ovens precooked meats, roasted coffee beans, and dried everything from flowers and cork to paper and potato chips.

Over time, changes were made in the microwave oven that made it useful in the home kitchen. Small ovens were built. Air-cooled magnetrons were made so no plumbing was needed to install the oven. Cheaper models were built.

Still, people were not buying the microwave oven. They were afraid food cooked in a microwave was harmful to eat. They thought microwaves could hurt them or make them sick. They found that microwave ovens did not brown foods or make them crisp.

Slowly, people came to know that microwaves are just like radio waves and light waves; they do not change molecules. They just move them around quickly to create heat. So microwaves do not harm food. They have not been shown to cause illness, either. And when a special kind of paper is wrapped around a food, microwaves can crisp and brown foods.

Home cooks finally began to trust microwave ovens. By 1975 more microwave ovens than gas ranges were sold in America each year. By the next year, more people in this country owned microwave ovens than owned dishwashers. Today millions of people around the world cook foods in microwave ovens.

Percy Spencer's curiosity turned a melted candy bar into a remarkable invention. It also earned the self-taught engineer an honored place in the National Inventors Hall of Fame.

Name _____

Questions about
The Mystery of
the Melted Candy Bar

Answer the following questions in complete sentences.

1. Why might Percy Spencer's friends have thought Spencer would not be successful in life?

_____

_____

2. Why was Spencer considered a World War II war hero?

_____

_____

3. What made Spencer believe that microwaves might be useful in cooking foods?

_____

_____

4. In what businesses did the first Radaranges find use?

_____

_____

5. Why were the first Radaranges not practical for use in home kitchens?

_____

_____

6. Why were people afraid to cook food in microwave ovens at first?

_____

_____

7. How do microwaves cook food?

_____

_____

Vocabulary

A. Use context clues from the story and/or a dictionary to help you match the
following words from the story with their definitions.

_____ 1. an electromagnetic wave that acts
much like a radio or light wave

_____ 2. something used to determine the
position of a faraway object

_____ 3. legally secured as an original
idea or invention

_____ 4. a tube that produces microwaves

_____ 5. left without a living parent

_____ 6. small particles of matter

Word Box
a. magnetron
b. orphaned
c. microwave
d. radar
e. patented
f. molecules

B. Now use the words from above in sentences of your own.

1. _____

_____

2. _____

_____

3. _____

_____

4. _____

_____

5. _____

_____

6. _____

_____

Convenient Kitchens

The microwave oven makes it possible to grab a quick hot snack or cook an entire meal in little time. Other appliances help make chores in the kitchen easier too. Identify the following kitchen inventions by their descriptions.

1. This 1858 invention became necessary after the 1810 invention of the tin can.

   _____

2. The first version of this item used to mix batters and doughs was created by Rufus Eastman in 1885.

   _____

3. The 1886 version of this Josephine Cochrane invention was hand-operated. Today, electricity runs these dish-cleaning machines.

   _____

4. This electrically powered food cooler replaced the ice box in 1913.

   _____

5. The pop-up version of this bread cooker was invented by Charles Strite in 1919.

   _____

6. Stephen Poplawski put together this spinning blades food chopper in 1922. It is popular in modern kitchens where it is used to puree foods and blend smoothies.

   _____

7. In 1952 Russell Hobbs made this machine that shuts off automatically when a popular hot drink is done brewing.

   _____

Just for the Fun of It

Well, now, Lacey's Cross-Road, Alabama, isn't exactly the most exciting place to live, but it sure is pretty. It was pretty in 1932, the year Al Glover was born, even though America was in the middle of the Depression. It was still pretty in 1945. That's when Al had his brainstorm. It happened like this.

Al's folks owned a farm, and they had to work it pretty hard. Al, of course, worked on it too. He picked cotton, did his share of plowing, and stripped corn from their stalks to feed the livestock. His dad also made roofing shingles, and Al helped with that too.

Things were tough in rural Alabama in those days. The war had provided jobs for people in other places, but not here. Here, a lot of folks were out of work and had to scratch out a living as best they could. Still, Lacey's Cross-Road was peaceful, with time to sit on porch swings and think about stuff.

Now, in those days, Al had a weakness for model airplanes. He loved the feel of the soft balsa wood and the gentle shape of the wings. He loved controlling the planes, making them glide and swoop and climb straight up. He and his friends would spend hours coaxing their planes into daring stunts. It was tricky. Sixty years ago, model airplanes were about as simple as they could be. They were attached to a long piece of fishing line or piano wire and "flown" by maneuvering the line. The planes climbed only as high as the line would allow, and when they crashed, they crashed in a big way.

"You know?" said Al one afternoon as he scooped up the shattered pieces of his downed plane, "There has to be a way to control these planes better."

"Sure," said one of his friends. "We'll put a tiny pilot inside and…"

The boys roared with laughter.

"No. I'm serious," said Al. "If you let the line go loose, the plane goes into a dive, and you can't pull it out."

"So?"

"So I'm going to figure something out."

Now, Al's father was not too crazy about Al's hobby. He thought his son was wasting his time messing with those toy planes, and he did nothing to encourage Al. Whatever Al knew about flight and **aerodynamics** he had learned on his own. By trial and error he discovered what worked and what didn't.

Al saved the broken parts of his downed planes and pieced them together in different ways. He tried this and that. He drew up designs, built them, and test-flew them, searching for something that would work.

Al knew the solution to uncontrolled nose dives had to be in the wing design. Maybe he had to change the wings in some way.

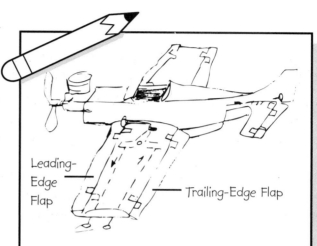

Leading-Edge Flap

Trailing-Edge Flap

Al fiddled and tinkered. He sketched, erased what he had sketched, and sketched again. And then one day, there it was in front of him, a totally new invention—the leading-edge flap.

Now, the year, if you remember, was 1945, and the world was at war. Never before had airplanes played such a major role in the fighting. While 13-year-old Al Glover was working on a wing design that would pull his balsa wood planes out of nose dives, Nazi engineers where puzzling over the very same problem. But they were using giant wind tunnels. All Al had was the warm Alabama breezes. The Germans had state-of-the-art machine shops, and there was Al, taping things together in his basement.

Planes of the time already had wings with a trailing-edge flap. This was a movable piece on the back edge of each wing that the pilot could move up or down from the cockpit. Moving the flap down made air flow over the top of the wing more quickly, increasing lift. Al's leading-edge flap had the same effect, producing even greater lift. Flaps enable planes to stay in the air at slower speeds. This is important during takeoff and landing.

It would certainly make a great ending to the story if Al had sent his sketches to the U.S. Patent Office and beat the German scientists to the punch. He didn't, though. In fact, it didn't occur to Al to try to market his invention until the 1950s. By then it was too late.

But the story has a good ending anyway. Al Glover became the inventor he was always destined to be. In the 1980s, while working at Chrysler, he won more patents than any other designer.

And what about all the work he put into his leading-edge flap? Al shrugs. "Aw," he says, "I just did it for the fun of it."

Name _____

Questions about
Just for the Fun of It

1. What were the model airplanes that Al Glover flew made of?

   ○ wire
   ○ papier-mâché
   ○ wax
   ○ balsa wood

2. How did Al come up with the idea for his leading-edge flap?

   ○ He copied a German design.
   ○ He discovered it through trial and error.
   ○ One of his friends suggested it.
   ○ His father helped him develop it.

3. Where on the airplane was Al's leading-edge flap located?

   ○ on the tail of the plane
   ○ on the nose of the plane
   ○ on the wing of the plane
   ○ on the belly of the plane

4. What did the leading-edge flap do?

   ○ It stopped the plane.
   ○ It allowed the plane to fly at high speeds.
   ○ It increased lift to keep the plane in the air.
   ○ It helped the plane fly straight.

5. What did Al become when he grew up?

   ○ an inventor
   ○ a stunt pilot
   ○ a race car driver
   ○ an airplane designer

Compound Words

A compound word is formed from two separate words. Some compound words
are joined together—sunglasses, keyboard, and suitcase. Some compound words
are not joined—first aid, high school, and fairy tale.

A. The words in the box can be joined to form four compound words that appear in the story and
four new ones. Write the eight compound words on the lines below. Circle the four words that
are found in the story. You may use a word more than once.

_____   _____

_____   _____

_____   _____

_____   _____

dives	pile	scatter	after
noon	storm	snow	nose
live	bleed	brain	stock

B. You can thank kids for two of these inventions. Read the clues and guess what they invented.
(Hint: They're all compound words.)

1. Chester Greenwood: "Maine winters are really cold, and my ears hurt so much, I had to
come up with something to keep them warm."

_____

2. John Hetrick: "My inflatable invention has saved a lot of lives in car crashes."

_____

3. Ralph Samuelson: "I invented a sport when I realized that snow was not the only thing you
could ski on." (Hint: There is a hyphen between the parts of this compound word.)

_____

Just for the Fun of It

Now It's Your Turn

Have you ever said, "You know what I think somebody should invent…?" Well, here's your chance. What would you invent if you had the chance? Here's how to do it:

Describe the purpose of your invention. How does it work? What does it do?

_____

_____

_____

_____

_____

Put it on the drawing board. Draw a sketch of your invention the way Al Glover did. Use arrows to point out the different parts. Then write the name of your invention on the line below the drawing board.

_____

Global Warming

Greenhouse gases in Earth's atmosphere trap heat, causing the planet to grow warmer.

Earth is getting warmer. About that, most experts agree. In the last century, the atmosphere of our planet has warmed by an amount between 0.5 and 0.9 degrees Celsius. This trend toward a warmer Earth is called global warming. Research points to **greenhouse gases** as one cause of global warming. These are gases that collect in the atmosphere. They work like the glass roof of a greenhouse, trapping heat near Earth's surface. Without them, Earth would be too cold. With too much of them, Earth would be too hot.

However, this is about all that experts agree on when it comes to the subject of global warming. Some scientists say that changes in Earth's temperature are part of a natural cycle of warming and cooling that has been going on for much of Earth's history. They point out that carbon dioxide and other greenhouse gases are produced naturally, and the levels of these gases in Earth's atmosphere have not been constant over time.

Many other scientists, however, feel that greenhouse gases produced by human activity are causing Earth's surface to warm much faster than it would naturally. Air conditioners, factories, and cars produce **carbon dioxide** and other greenhouse gases that enter the atmosphere. Most research shows that one hundred years from now, this planet will be between 1.0 and 3.5 degrees Celsius warmer. Some scientists feel there will be an even greater increase.

This rise in temperature could have dramatic effects. Ocean levels could rise. Some areas could have dramatic changes in **climate**. These changes would certainly affect the way people live. Some people might find the climate better for farming, and others might no longer be able to farm. It could become too hot or too damp. Some plants and animals would not survive the change. Many are very sensitive to temperature. Animals that migrate might have to go much farther to find a climate that suits them. Some might not be able to make the journey.

Read and Understand, Science • Grades 4–6 • EMC 3305

If we could reduce the amount of greenhouse gases in the atmosphere by simply turning off our air conditioners and cars, global warming might be less of a problem. Unfortunately, greenhouse gases last a long time. Even if we could turn everything off, it would take dozens of years for the gases to disappear. This is why many scientists want people to start making changes now.

Solar energy, that is, energy from the Sun, may be part of the solution. The main sources for energy used in the world today are **fossil fuels**—coal, oil, and gas. The burning of these fuels gives off a lot of carbon dioxide. Energy from the Sun gives off no gases. It does not have to be mined from the earth. It does not pollute air, ground, or water. Still, solar technology is fairly new. It is still more expensive to use solar energy than fossil fuels. Once scientists come up with cheaper ways to use solar energy, we may be able to reduce the amount of fossil fuels we use.

The other part of the solution involves you and people like you. If more people commit to saving energy, it will make a big difference. You can walk to where you need to go, instead of riding in a car or on a bus. You can talk to adults about planting trees around your house or school. This will help cut the amount of energy used to cool the buildings in summer. You can recycle, and choose products that have little packaging. This means less energy will be used to collect and dispose of garbage. Most importantly, you can talk to others, making people aware of the problem and solutions.

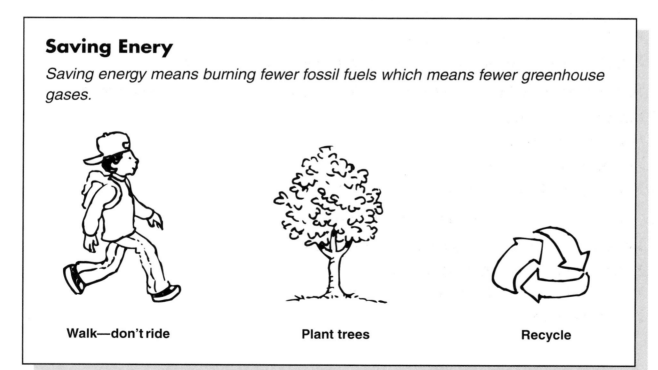

Saving Enery

Saving energy means burning fewer fossil fuels which means fewer greenhouse gases.

Walk—don't ride   **Plant trees**   **Recycle**

Questions about
Global Warming

Fill in the bubble that best completes each sentence or answers each question.

1. In the last 100 years, Earth's atmosphere has _____.

   ○ become cooler
   ○ stayed exactly the same temperature
   ○ become warmer
   ○ never been measured

2. Greenhouse gases work like the glass roof of a greenhouse to _____.

   ○ trap water on Earth's surface
   ○ let out extra heat
   ○ trap heat near Earth's surface
   ○ keep out some of the rain

3. How do many scientists believe that people contribute to global warming?

   ○ People use up too much of Earth's oxygen when they breathe.
   ○ Air conditioners, cars, and factories produce greenhouse gases.
   ○ Machines and cars heat up the air.
   ○ People are turning off their air conditioners, making Earth warmer.

4. Which of these is **not** described in the article as a possible effect of global warming?

   ○ Climates would change.
   ○ Ocean levels could rise.
   ○ Plants would be smaller.
   ○ Some animals would not survive the change.

5. What can you do to help prevent global warming?

   ○ Recycle and choose products with little packaging.
   ○ Walk instead of taking a car or bus.
   ○ Talk to others about the problem and solutions.
   ○ all of the above

Name _____

Vocabulary

Use clues in the article or a dictionary to help you match each word or phrase with its meaning.

Word Box				
technology	climate	cycle	global warming	pollute
fossil fuels	atmosphere	migrate	solar energy	carbon dioxide
constant	research	mined	dozen	

1. to move from one region to another _____

2. the application of scientific knowledge to solve a problem _____

3. one type of greenhouse gas that is produced by people _____

4. the slow rise in temperature of the air that surrounds Earth _____

5. a series of events that is repeated again and again _____

6. twelve _____

7. always the same _____

8. the usual weather for an area _____

9. energy from the Sun _____

10. the air that surrounds Earth _____

11. coal, oil, and gas _____

12. to make something dirty or mix in things that don't belong _____

13. dug out of the ground _____

14. careful study and the learning that follows _____

Name _____

Using a Table of Contents

Read the Table of Contents and use the information to answer the questions below.

1. In which chapter could you read about how scientists learned that Earth is getting warmer?

   _____

2. If you wanted to learn what might cause global warming, what pages would you read?

   _____

3. Which chapter would you read to learn what Earth's atmosphere is made of?

   _____

4. If you would like to help prevent increased global warming, on what page might you begin reading?

   _____

5. How can a Table of Contents help you?

   _____

   _____

Here's Looking at You, Kid

Long, long ago, when Earth was still quite young, life was very simple. Most of it, in fact, was **microscopic**, floating invisibly in the newly formed oceans. And yet these primitive creatures still knew a little something about the world around them. They knew light and dark, and they could sense warmth and cold, thanks to something called **eyespots**, small clumps of cells that are sensitive to light. The eyespots worked well enough for the first life-forms, but they would never do for the more complex animals that were to come. Something much better would be needed.

Some primitive animals like the planarian still have eyespots. The planarian is a very tiny flatworm with a special talent. If it is cut into pieces, each piece will grow into another planarian!

eyes

The scallop's eyes are arranged in rows. The scallop can't see objects very clearly, but it can tell when something's moving.

The first big improvement came when eyes developed the ability to detect motion. This would turn out to be very important. It would help an animal catch food and avoid everybody else who thought *it* was food.

All animals' eyes are special and unique because each species has different needs. Animals that hunt other animals are called **predators**. Just being able to see motion is not enough. Predators must also be able to judge distance. Predatory birds, like the eagle, watch for food as they soar high above the ground. That's why birds have the best distance vision of any creatures on Earth.

But what about the animals that are hunted? What about the **prey**? These three animals are all prey to other animals, but they're not helpless. Look at their eyes. Where are they?

 **Zebra**    **Rabbit**    **Mouse**

Notice that their eyes are on either side of their head. This is the perfect position because it lets an animal see if something is sneaking up behind it.

Animals' eyes can be located in some very wacky places—at least that's what we might think. But to a hermit crab, having your eyes at the ends of long stalks is a good idea indeed. The hermit crab has a very soft body but no shell to protect it. So it scrounges around on the beach, looking for shells that other animals have left behind. Very often these shells are huge and would block much of the hermit crab's vision if it weren't for nature's version of the periscope. The crab's stalk eyes clear the shell and allow the crab to see all around it.

Many animals, like the raccoon and skunk, are **nocturnal**. Nocturnal animals are active at night. When the sun sets and the land grows dark, nocturnal animals find themselves in a world with very little color. So if you think that nocturnal animals have poor color vision, you're right. They don't need it. Raccoons can't really see anything brighter than the color green. Red, yellow, and orange just look dark to them. But what nocturnal animals do need is good night vision. (But you probably figured that out already.)

The crocodile's eyes are on top of its head, and this, too, is a good idea. The sneaky crocodile can swim just beneath the water's surface and still see what's on top that might be good for lunch.

Animals have eyes that allow them to gather the information they need to survive in their environment. For some species, this means being able to escape predators. For others, survival means being able to catch prey animals. In either case, the eyes of the animal help it to accomplish its goals.

Questions about
Here's Looking at You, Kid

Fill in the bubble that best completes each sentence or answers each question.

1. What are eyespots?

   ○ markings on an animal that look like eyes

   ○ organs that can detect movement

   ○ a cluster of cells that can detect light and dark

   ○ very small eyes

2. How does having eyes on either side of its head benefit an animal?

   ○ It gives the animal better distance vision.

   ○ It allows the animal to see better in the dark.

   ○ It gives the animal three-dimensional vision.

   ○ It helps the animal to see if something is sneaking up on it.

3. The eyes of predators must be able to _____.

   ○ judge distance and see motion

   ○ see color and sense heat

   ○ blink quickly and see clearly underwater

   ○ glow in the dark and change their size

4. What do nocturnal animals need most?

   ○ good color vision

   ○ the ability to see well in the dark

   ○ good distance vision

   ○ eyes that reflect sunlight

5. How does having eyes at the ends of long stalks help a hermit crab?

   ○ The eyes look weird and scare away enemies.

   ○ Eyes on stalks are better for seeing in the dark.

   ○ Vision will not be blocked by a large shell.

   ○ Having "periscope" eyes helps to attract a mate.

Vocabulary

Name _____

Here's Looking at You, Kid

A. Synonyms are words that have **similar** meanings. Find a synonym in the Word Box for each of these words from the story. Look out! Some words in the box will be left over.

1. primitive _____

2. detect _____

3. motion _____

4. judge _____

5. version _____

Word Box

form	side
complex	movement
visible	simple
determine	sense

B. Antonyms are words that have **opposite**, or nearly opposite, meanings. Unscramble these words to find an antonym for each word on the left.

1. primitive _____

2. predators _____

3. brighter _____

4. unique _____

Word Box

mpocexl

ryep

ledurl

dyrronia

C. Sometimes, the prefix *in-* means "not," so it can turn many words into antonyms. For example, *inactive* means "not active." If you remove the prefix, you get the word *active,* an antonym for *inactive*. Circle all the words whose meaning turns into just the opposite by adding the prefix *in-.*

dependent   nocturnal   sensitive

detect   visible   dangerous

©2002 by Evan-Moor Corp.   113   Read and Understand, Science • Grades 4–6 • EMC 3305

Here's Looking at You, Kid

Over the years, Hollywood has created a lot of aliens. Now it's your turn.

Planet X-14-006 circles the third star in the constellation Bootes. The planet receives only 6 hours of sunlight each day. (A day lasts for 45 Earth hours.) Everything on the planet is underwater with one exception: a very tall mountain rises 6,000 feet (1,829 m) from the ocean. There is life in both the ocean and on the mountain.

Design a water animal and a land animal for the planet, and describe what kind of eyes each animal would have to successfully survive on the planet.

Sketch your animals in the boxes. On the lines below the sketches, describe the animals' eyes.

water animal

_____

land animal

_____

Understanding Extinction

Extinctions have occurred throughout history. In one short period, millions of years ago, 90 percent of all **marine species** died out. Dinosaurs disappeared too. All that remain of these ancient life-forms are **fossils** and bones.

There are many reasons ancient extinctions happened. Natural processes of **competition** and **evolution** were probably responsible for most extinctions. Climate changes and **natural disasters** likely killed some animals too. One **theory** even suggests dinosaurs died when a comet hit the Earth.

Today human activity puts many animals in danger of extinction. Cutting down trees in rainforests puts animals at risk. Collecting **natural resources** destroys animals' homes. So does turning wetlands into farmland. Man-made pollution kills animals and plants. Hunting (for food, sport, and skins) endangers the survival of some animal species.

Saving animal species is not as easy as it sounds. If trees in the rainforest are not cut down, some people will go without shelter or heat. If oil and metals are not mined, modern life will stand still. Even animal hunting is not a clear-cut issue. Some people believe humans have the responsibility to manage wildlife because we have changed their world. They say some animals must be killed to keep the ratio of predators and prey in balance. They also argue that legal hunting concessions keep illegal hunting to a minimum.

Read the advertisement below encouraging tourists to visit the African wild. Then read the letter written in response to the ad. Based on what you read, decide for yourself how best to save animals from extinction.

African Adventures Safari Camp Introduces.... Mix and Match Safaris

Spend your next vacation in the African wild! At *African Adventures Safari Camp* you will sleep in clean cabins in the African jungle. Elephant and lion sounds will wake you each morning. After breakfast a safari guide will lead you to the trophy you hope to take home. Or perhaps he will show you the best places to take pictures of wild beasts. It's up to you! Mix and match elements from any of our safari adventures to meet your vacation dreams.

- Picture-perfect Adventure—Our guides will lead you through untamed jungles on horseback or foot. Stop often to take pictures of giraffes, buffalo, and hippopotami.

- Wonderful Water Adventure—Scuba dive, canoe, or raft your way along the coast of Africa. Tropical fish swimming in the Indian Ocean will display their colors to you. Lions prancing along the shore will show you their **stalking** dance.

- Hunters' Paradise Adventure—Allow our professional hunters to lead you to a once-in-a-lifetime catch! For only $250 you can kill a warthog. Or spend a little more and take home an elephant head ($4,500) or a lion skin ($15,000).

Dear African Adventures Safari Camp,

I just read your advertisement for "mix and match" safari adventures. I know that some African communities rely on tourism for money. However, I am not happy that companies like yours are exploiting the continent. I find it appalling that anyone who can fork out enough dough can kill an elephant or a lion. Trophy hunting is one of the reasons so many animals are on the endangered species list. You may consider your game reserves a "fair" place to kill animals. After all, your safari camp has a license to do the business it does. Yet an animal that is shot doesn't know the difference between the "legal" hunters from your camp and a common **poacher**. Some causes of **animal endangerment** are hard to prevent. Climatic changes are out of our control. Decisions about whether to make farmlands available to starving people at the expense of saving the rainforest are hard to make. Trophy hunting, though, is an easy thing to stop.

Your ad makes it sound harmless to vacation in the wild. It is not harmless. Even your guests who don't hunt are intruding on a world that belongs to animals and African natives. Vacationers pollute the wild with noise and garbage. When an animal threatens the safety of a safari camp guest, the animal is killed. When an animal destroys camp property, the animal is killed. Why shouldn't a wild animal threaten guests and destroy property? You have set up camp in the middle of its home!

Your advertisement brags of the "untamed jungles" around your camp. How long do you think the wild will remain unspoiled if you keep welcoming humans into its front door?

Signed,

Angry in Alabama

Questions about
Understanding Extinction

Fill in the bubble that best answers each question.

1. Which statement about extinctions is **not** true?

   ○ Extinctions have occurred throughout history.

   ○ Human actions have caused all animal extinctions.

   ○ People sometimes destroy animal homes when they drill for oil or cut down trees.

   ○ Some types of pollution kill plants and animals.

2. What is the *African Adventure Safari Camp's* opinion of vacationing in the African wild?

   ○ Vacationing in the African wild is an adventure.

   ○ Vacationing in the African wild is bad for the land, people, and animals of Africa.

   ○ Vacationing in the African wild is too expensive.

   ○ Vacationing in the African wild is scary and unsafe.

3. What is "Angry in Alabama's" opinion of vacationing in the African wild?

   ○ Vacationing in the African wild is exciting for everyone.

   ○ Vacationing in the African wild is a fun experience for hunters.

   ○ Vacationing in the African wild is too expensive.

   ○ Vacationing in the African wild exploits animals and Native African cultures.

4. What is meant by the idiom "fork out enough dough"?

   ○ travel a long distance

   ○ serve an excellent pie

   ○ spend enough money

   ○ complain enough

5. Which of the following is a cause of extinctions?

   ○ cutting down trees in rainforests

   ○ manmade pollution

   ○ natural processes of competition and evolution

   ○ all of the above

Name _____

Vocabulary

Words from the story are defined below. Read the definitions. Then complete each of the sentences with the correct vocabulary word.

appalling—horrifying and shocking

climate—average weather conditions of a certain place over a period of years

competition—a situation in which two animal species fight for the same resources

marvel—to view with wonder and amazement

evolution—the theory that all species develop and change over time

extinction—no longer existing

hunting concession—land on which a country's government permits hunting

intruding—forcing oneself upon something or someone

tourism—the business of providing for tourists, people who travel for pleasure

natural disasters—natural occurrences that cause great damage such as a flood or tornado

natural resources—things from nature that are useful to humans such as metals and oil

predator—an animal that kills other animals for food

poacher—someone who kills animals illegally

trophy—a lion's skin, deer's head, or other evidence of one's hunting success

1. The winter _____ in Montana is snowy and cold.

2. Plants and animals on the endangered species list are in danger of _____.

3. Earthquakes and volcanoes are frightening _____ _____.

4. Coal and gold are valuable _____ _____.

5. Although the _____ had no hunting license, he killed a deer.

6. The theory of _____ suggests species have changed over time.

7. Many hunters mount the horns or antlers of an animal they kill as a _____.

8. The jobs of many people in the small beach community depended on _____.

9. When you visit the Grand Canyon, you will _____ at the enormous size of it.

10. The orca, or killer whale, is a fierce _____ that hunts in packs.

Name _____

Reading a Map

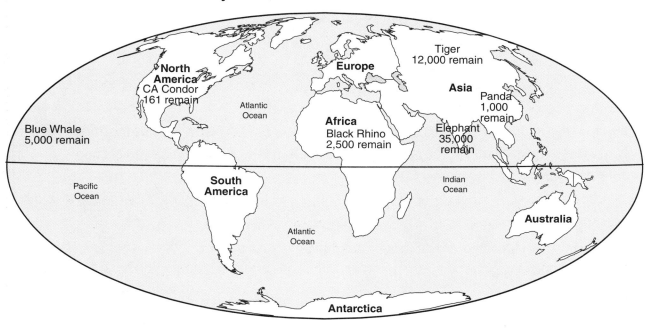

Thousands of animals are at risk of extinction. Critically endangered species are in need of immediate human help to survive. Threatened species are declining in numbers. Rare species are at risk, but not in immediate danger of extinction.

Read the map below to help you see where some at-risk animals live. Then answer the questions that follow.

Primary Homes of Some At-Risk Animals

1. Where do the 1,000 remaining pandas live?   _____

2. How many black rhinos remain in the world?   _____

3. Where do most of the world's remaining Asian elephants live? _____

4. Which endangered animal lives in both the Pacific and
   Atlantic Oceans?   _____

5. Which animal shares a home with the Asian elephant?   _____

6. Which country is home to the 161 remaining California condors? _____

©2002 by Evan-Moor Corp.   119   Read and Understand, Science • Grades 4–6 • EMC 3305

McDonald Observatory

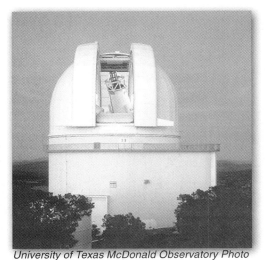

University of Texas McDonald Observatory Photo

McDonald Observatory sits atop the mountains of west Texas. The air is clear and there are no major cities to "pollute" the night sky with light.

In the mountains of west Texas, under a black sky salted with light from thousands of stars, stands a group of white and silver domes. These domes house the mighty telescopes of McDonald Observatory.

The site is ideal for stargazing. The domes of the observatory are set on mountaintops. At about 7,000 feet, the air is dry and clear. Few people live in the area, so the sky is dark. In the city, where there are lots of lights, stars are hard to see. This is what astronomers call "light pollution."

In 1939 work on the first big telescope at the observatory was completed. This was the Otto Struve telescope. It has seen a lot of use since then. In fact it has been used every clear night since it was finished. It is often used for public viewing of the stars.

The Otto Struve telescope is a reflector telescope. This means it uses mirrors to collect light from things in space. In this way, distant objects look brighter and closer. All of the big telescopes at McDonald Observatory are reflector telescopes. When it was built, the Otto Struve telescope was the second largest in the world. Its mirror is more than 2 meters (82 inches) across. That's as wide as many professional basketball players are tall!

The Harlan-Smith telescope is even bigger. This telescope weighs about 190 tons. Its mirror is nearly 3 meters (almost 9 feet) across. NASA has used the Harlan-Smith telescope to get ready for missions in space. They learned about the atmospheres of distant planets such as Jupiter. They learned how far away these planets are. They used information gathered with the telescopes to help them decide what they wanted to learn about the planets. It's a lot like using a guidebook to learn about a place before you visit there on vacation. The Harlan-Smith telescope was used before the Viking mission to Mars. It was also used to prepare for Voyager missions, which explored the outer parts of our solar system.

In 1998 the Hobby-Eberly telescope was finished. It is the third largest telescope of its kind in the world. Astronomers use it to study what stars are made of and how they move. They learn about how hot stars are and how far away they are. What they learn helps them understand how stars are born and how they die. It is a powerful tool.

Other things are done at the observatory as well. The McDonald Lunar Ranging Station shoots laser beams at the Moon. When the Apollo astronauts went to the Moon in the late 1960s and early 1970s, they left reflectors there. A 30-inch telescope at McDonald Observatory shoots laser beams at these reflectors. Then the ranging station measures how long it takes the laser beams to bounce off the reflectors and return to Earth. These numbers allow scientists to measure any movement of the Moon, down to the centimeter. The ranging station also measures movement of satellites.

University of Texas McDonald Observatory Photo

Astronomers don't usually look through the Harlan-Smith telescope (shown here) or the Hobby-Eberly telescope. They place instruments at the back of the telescope. These instruments take pictures, measure brightness, and analyze light.

There was a time when astronomers learned about space mainly by gazing through telescopes. Today astronomy is different. Scientists attach cameras and computers to telescopes. They use special devices to study light from objects in space. The information they gather tells them a great deal about comets, stars, planets, and even galaxies.

As early as 1610, a man named Galileo Galilei used a telescope to study the stars. Galileo built and used refractor telescopes. These telescopes used curved lenses to collect light, rather than the mirrors used in today's telescopes. With a telescope he made himself, Galileo discovered the moons of Jupiter in 1610.

Galileo Galilei
1564–1642

Name _____

Questions about
McDonald Observatory

Answer the following questions in complete sentences.

1. Why is McDonald Observatory's location in the mountains of west Texas ideal for stargazing?

_____

_____

_____

2. How does a reflector telescope make distant objects seem brighter and closer?

_____

_____

_____

3. Which of the big telescopes at McDonald Observatory did scientists at NASA use to help them prepare for the Viking mission to Mars?

_____

_____

   Describe three ways that this telescope helped NASA scientists prepare for missions in space.

_____

_____

_____

_____

4. Describe how the McDonald Lunar Ranging Station is used.

_____

_____

_____

_____

Name _____

Vocabulary

Complete the puzzle using words related to the article. Use a dictionary to check the meanings of your answers, if needed.

Across

2. a concentrated beam of light; letters stand for "Light Amplification by Stimulated Emission of Radiation"
3. a curved glass that makes things look larger or closer than they are
5. the astronomer who discovered moons around Jupiter in 1610
6. a building or group of buildings that house telescopes for viewing the stars
9. it looks like a star with a shining tail
10. a device that makes objects in space seem brighter and closer than they are
13. often means an object made by people and sent into orbit around Earth
14. the study of things and events in space

Down

1. devices; tools
2. illumination that makes it possible to see
4. a name assigned to U.S. missions to explore outer reaches of our solar system
7. a place; location
8. with 2 down, it means light from buildings and streets that makes it hard to see the stars
10. 2,000 pounds or 907.2 kilograms
11. a name of a U.S. mission to Mars
12. the U.S. agency in charge of exploring space; letters stand for "National Aeronautics and Space Administration"
13. from Earth, it looks like a twinkling point of light in the night sky

Name _____

Using a Book Index

Here is part of an index from a book. Read the index and use the information to answer the questions that follow.

1. On what pages would you look to find out how hot a star is?

   _____

2. What page could you turn to in order to read about pulsars?

   _____

3. Where would you look to learn the names of the planets in our solar system?

   _____

4. Why is "space station" listed after "space exploration"?

   _____

5. What would you learn about if you read pages 75–77?

   _____

6. How can a book index help you?

   _____

   _____

Cells: Structure and Function

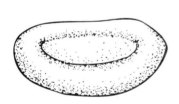

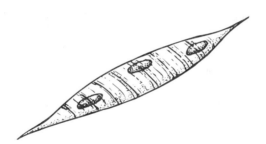

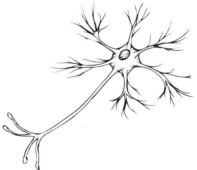

Red blood cells are round.

Muscles cells can change shape.

Nerve cells have long arms that send messages to other nerve cells.

Every living thing is made up of one or more **cells**. Many life-forms, such as bacteria, are made of only one cell. Your body is made of more than ten trillion cells.

Cells are the basic unit of life. Each cell contains all the materials needed for life. In one-celled **organisms**, the cell must perform all the jobs needed to keep the organism alive. In many-celled organisms, different cells have different jobs. Together, all the cells work to keep the organism alive.

To look at a cell, you would need a microscope. That's because cells are microscopic, or too small to see without a microscope. If you were to look at several cells from your body under a microscope, you would see that cells do not all look the same.

The structure, or shape, of a cell relates to its function. **Skin cells** are flat. This allows them to pack together in layers to create a tough barrier between the environment and your body's inner parts. **Nerve cells** are specially designed to send and receive messages. Long arms extend from each cell and reach for other nerve cells. Messages move quickly from one cell to the next, delivering important information. Cells that make up your muscles are able to shorten or lengthen, producing movement. Blood cells are rounded so that they can move easily through small blood vessels.

Each kind of cell in your body has a particular job to do. **Red blood cells** carry oxygen to all other cells in the body. **White blood cells** fight invading germs. **Muscle cells** are responsible for movement, and **fat cells** are in charge of storing energy. Each of your cells is specially designed to carry out its job.

Just as your body is made of many parts, so is each cell. And just as each cell is specially made for a particular job, so is each cell part specially designed to carry out its particular function.

Each cell is enclosed by a **cell membrane**. The cell membrane holds the inner parts of the cell together. Because one of its functions is to allow materials to flow in and out of the cell, the cell membrane has small holes, or **pores**. Plant cells have another layer around the cell membrane called a **cell wall**. The cell wall's job is to support the plant, so it is made of stiff materials.

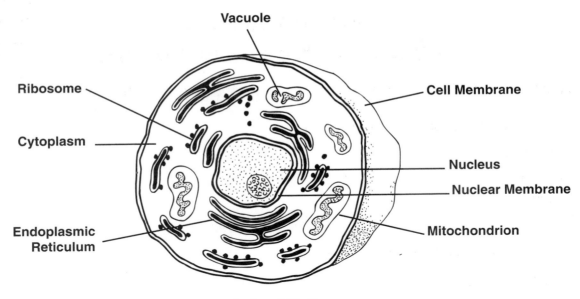

Vacuole

Ribosome

Cytoplasm

Endoplasmic Reticulum

Cell Membrane

Nucleus

Nuclear Membrane

Mitochondrion

Animal Cell

The **nucleus** is the cell's command center. It contains DNA—the cell's **genetic material**. DNA stores information about the cell and directs the type of work that the cell does. It also contains a sort of "master plan" for the cell, so that each generation of cells is just like the generation before. Because the nucleus contains such important information, it is separated from the rest of the cell by its own membrane, called the **nuclear membrane**.

The inside of the cell is filled with **cytoplasm**. Cytoplasm is a watery gel. Its structure is perfectly suited to its job—to support the cell's other parts, called organelles. Some of these **organelles** include ribosomes, mitochondria, vacuoles, and the endoplasmic reticulum.

Ribosomes make proteins for the cell. Proteins are needed for cell growth and repair. Ribosomes are small and compact and are located in several different places within the cell. This way, the cell is able to make proteins in more than one location within the cell.

Mitochondria have the job of converting the energy in food into a form the cell can use to grow, divide, and do its work. The inner membranes of mitochondria have many folds. Energy is converted on the surfaces of these folds. Having many folds allows the mitochondria to convert a lot of energy.

Vacuoles are like tiny storage tanks, holding and releasing materials that the cell needs. Vacuoles can shrink or expand, depending on how much material they are holding at the time.

The **endoplasmic reticulum** is a web of tubes and pouches that extends throughout most of the cell. This web acts as a delivery system as materials are able to travel through the tubes all around the cell.

Both cells and cell parts are specially designed to carry out a particular job, or function. The structure of each is perfectly designed to carry out that function.

Name _____

Questions about
Cells: Structure and Function

Fill in the bubble that best completes each sentence or answers each question.

1. _____ are the basic units of life.

   ○ Organisms
   ○ Functions
   ○ Cells
   ○ Nerves

2. Every cell's structure helps it _____.

   ○ support the plant
   ○ move
   ○ fight off other cells
   ○ do its job

3. Why is a blood cell rounded?

   ○ so that it can move easily through small blood vessels
   ○ so that every side will be the same
   ○ to protect the nucleus
   ○ in order to take in food

4. What encloses each cell?

   ○ the nuclear membrane
   ○ the cell membrane
   ○ the nucleus
   ○ the endoplasmic reticulum

5. The cell's command center is called _____.

   ○ mitochondria
   ○ cytoplasm
   ○ the cell membrane
   ○ the nucleus

Vocabulary

Use the words in the Word Box to complete the sentences below.

Word Box					
producing	microscope	extend	organisms	particular	barrier
designed	generation	gel	convert	responsible	

1. I have red hair, as do my uncle and my grandmother. At least one person in every

   _____ of my family has it.

2. Some _____ are so small that you can't see them without

   a _____ .

3. This special camping chair can _____ into a cot for sleeping.

4. When my parents go out, I am _____ for taking care of my younger
   sister.

5. Let's put up this tarp, so we'll have a _____ against the sun.

6. The wind blew through the eaves, _____ a moaning sound.

7. A _____ is a thick substance. It is firmer than a liquid, yet less firm
   than a solid.

8. Every one of us has a _____ place that we like to sit.

9. This van is _____ to seat no more than seven people.

10. Wires _____ from the telephone pole to our house.

Name _____

Cells: Structure and Function

Study the diagram of an animal cell below. Label each cell part and tell what it does. The cell membrane and nucleus are done for you.

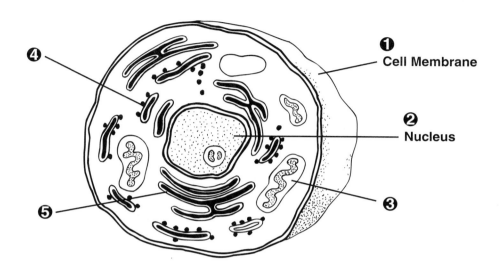

❶ Cell Membrane

❷ Nucleus

❸

❹

❺

1. **Cell Membrane**

   The membrane holds the inner parts of the cell together and allows materials to flow in and out of the cell.

2. **Nucleus**

   The nucleus directs the type of work that the cell does. It also contains DNA.

3. _____

   _____

4. _____

   _____

5. _____

   _____

Crooked Cells

Nellie Thomas cooed as only a mother can, trying desperately to still the cries of her screaming infant. "It's okay. Mama's here," she whispered, as if the tenderness of her voice would be enough to calm him. But baby George, just a few weeks old, wailed on, wracked by a terrible pain. Nellie rocked him through the long night, not knowing why his arms and legs were so stiff or why he couldn't stop crying. Helpless, all she could give was her love.

The following day, Nellie consulted the local doctor. But to her dismay, he frowned sadly and shook his head. "I'm sorry, Nellie. I don't know what's wrong with George."

In the months that followed, Nellie's baby had many episodes of the same strange symptoms. Each time, he suffered intense pain and stiffness so severe that he couldn't even bend his arms or legs. Nellie often had to rush him to the hospital, but even there, the doctors couldn't come up with a diagnosis. They all agreed on one thing, though. Nellie Thomas's child probably wouldn't live long.

It was five years before George Thomas's illness was finally identified. A doctor who had come to town for a medical conference explained to Nellie that her son was suffering from sickle-cell anemia, a disease that affects the body's red blood cells.

Normal red blood cells look like little puffy disks with their centers punched in. They are soft and flexible so they can squeeze through tiny blood vessels, called capillaries, which can sometimes be smaller than they are.

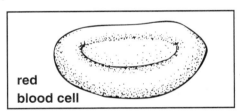

red
blood cell

The job of the red blood cells is to pick up oxygen in the lungs and transport it to the other cells in the body. They can do this because they contain a substance called hemoglobin, which easily binds with the oxygen. But in someone who has sickle-cell anemia, a mistake has occurred in the instructions to make hemoglobin. The faulty hemoglobin molecule can still bind with oxygen, but when the red blood cells release the oxygen, the hemoglobin molecules have a tendency to join together and form stiff fibers. The fibers distort the shape of the red blood cell. Instead of looking like little round disks, they turn into something that looks like the picture to the right.

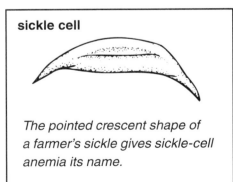

sickle cell

The pointed crescent shape of a farmer's sickle gives sickle-cell anemia its name.

Because of their weird shape, sickle cells often get stuck as they try to move through narrow blood vessels. Their sharp edges catch and stick, and soon there is a cell pileup, like a multiple car crash on the highway. The pileup blocks the way and prevents the blood from passing. This causes very painful swelling, mostly in the hands and feet. But blood vessels anywhere in the body can become blocked—even in the heart and lungs, and even in the blood vessels leading to the brain.

Sickle-cell anemia is caused by an error in the instruction manual found in each of our body's cells. The error, called a mutation, occurs in the gene that tells the body how to make hemoglobin. A person with sickle-cell anemia inherits the disease from his or her parents. In other words, the disease is passed from parent to child.

George Thomas was diagnosed in the 1950s, when the only treatment for sickle-cell anemia was blood transfusions. George was so sick he didn't even start school until he was seven. Nobody thought he'd be able to keep up, but keep up he did, and he graduated from high school with honors. Sadly, however, the disease took its toll, and George died at the age of forty.

Rabiyyah is George's grandniece. She, too, has inherited sickle-cell anemia. Scientists were able to tell Rabiyyah's parents their chances of having a child with sickle-cell anemia before they decided to start a family. This information helped Rabiyyah's parents make important decisions. It also helped them prepare for the special care Rabiyyah would need once she was born.

Many sufferers of sickle-cell anemia are excited about the advances being made in the field of genetic engineering. One day soon, scientists may actually be able to remove the mutation in the hemoglobin gene of sickle-cell anemia patients. Genetic engineering raises some ethical questions that must be discussed by people living in a society. One thing is for sure—the more we know about how a disease works, the better we'll be able to fight it.

Questions about Crooked Cells

Fill in the bubble in front of the answer that correctly completes each sentence.

1. Red blood cells are flexible so that _____.

   ○ they can carry oxygen
   ○ they can pass through capillaries
   ○ they can form a sickle shape
   ○ they can stick together

2. The substance in red blood cells that binds with oxygen is _____.

   ○ sickle-cell anemia
   ○ a gene
   ○ a stiff fiber
   ○ hemoglobin

3. Genetic engineering may enable scientists to _____.

   ○ make sickle cells round
   ○ end all disease
   ○ fix mutated genes
   ○ make hemoglobin carry more oxygen

4. Sickle-cell anemia is _____.

   ○ an inherited disease
   ○ caused by the mutation of a gene
   ○ a disease that affects red blood cells
   ○ all of the above

5. Sickle-cell anemia can be painful because _____.

   ○ blood vessels become blocked
   ○ red blood cells crash together
   ○ sickle cells get too big
   ○ there is no oxygen in the blood

Crooked Cells

Vocabulary

Use these words from the story to complete the crossword puzzle.

Word Box						
symptoms	faulty	severe	suffered	distort	episodes	diagnosis
gene	inherited	ethical	mutation	fibers	dismay	identified

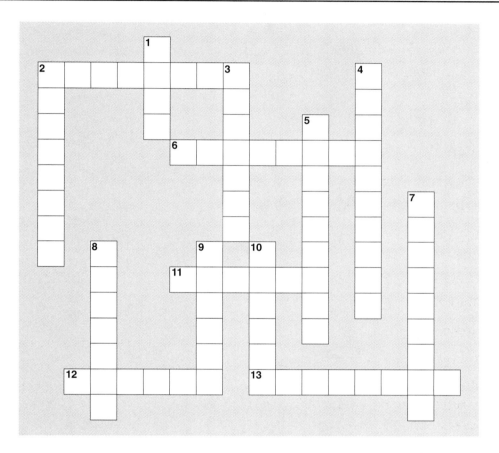

Across

2. felt pain
6. a change in genetic material
11. thread-like strands or structures
12. defective; imperfect
13. happenings; separate events

Down

1. a structure that tells your body how to form and grow
2. signs of illness
3. to twist out of shape
4. recognized; named
5. identification of a disease
7. received from birth from one's parents
8. having to do with right and wrong
9. a feeling of alarm or disappointment
10. extreme; serious

Name _____

Many Kinds of Cells

Each type of tissue in your body is made of a specific kind of cell. Use the descriptions below of some different types of cells to identify the picture of the cell. Write the number of the type of cell in the box beside its picture.

1. Skin Cells

Skin cells are flat. This allows them to pack together in layers to create a tough barrier between the environment and your body's inner parts.

2. Skeletal Muscle Cells

Muscles cells, which are responsible for movement, can change shape. Skeletal muscle cells are striped and contain several nuclei.

3. Nerve Cells

Nerve cells are specially designed to send and receive messages. Fibers extend from each cell and reach for other nerve cells. Small extensions receive electrical signals and channel them to the cell.

4. Red Blood Cells

Red blood cells look like puffy disks with their centers punched in.

5. White Blood Cells

White blood cells may be twice as large as red blood cells.

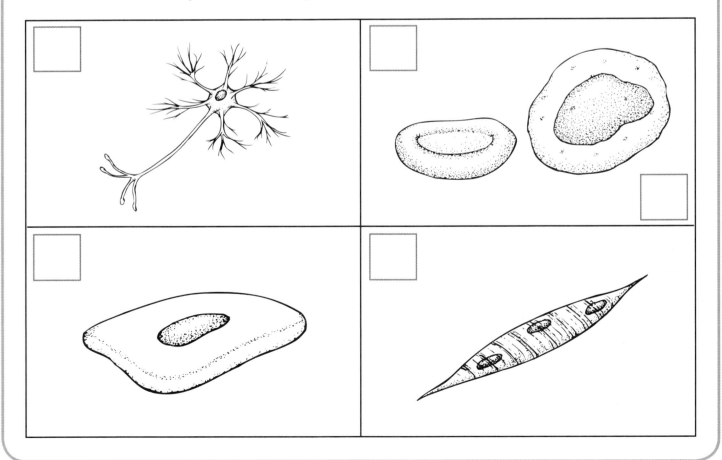

Communicating Through the Ages

Think about all the things you have done today that have involved information exchange. Did you read a book or watch TV? Did you e-mail or call a friend? Our communication options have changed a lot over the years. Read the time line to learn how people have shared information through time, from the invention of the printing press to the Internet.

1449 Johann Gutenberg invents a printing press that uses movable type. He places metal letters on a tray to form words. He then uses a wine press and ink to print the words on paper. He repeats the steps to make new copies. Before the printing press, monks took years to copy books by hand with quill pens. Gutenberg's machine gives even common people access to books and learning.

1564 A soft, black mineral—good for writing and drawing—is found in England. People hold chunks of it in their hands to write. Then they begin to wind it in string or cover it with leather to keep their hands clean. Finally, the graphite is placed in a hole drilled into a stick of wood. The pencil is born. It provides people with an erasable writing tool.

1837 Samuel Morse, a painter turned inventor, learns that pulses of electric current can be sent over a wire. He writes a code of dots and dashes. The telegraph allows people to send messages over a long distance.

1868 Latham Sholes patents a machine with letters on keys. Many before him have tried to build a writing machine. Sholes' typewriter is the first to type faster than a person can write. Improvements later lead to the electric typewriter and eventually the computer word processor.

1876 Alexander Graham Bell patents the telephone. Bell has been studying sound since childhood. He is interested in the work others do with sound. One time he misinterprets a German book. The writer talks about sounds that can be made with a piece of metal. Bell translates the reading to mean sound can be transmitted over wire. Trying to achieve this, Bell invents the telephone.

1877 Thomas Edison invents the phonograph. He mounts tin foil on a grooved cylinder. He attaches a needle and a little metal cup to the cylinder. Edison says, "Mary had a little lamb" into a mouthpiece attached to the metal cup. Vibrations from his voice reach the cup and move the needle. The needle etches the foil. The phonograph records Edison's words. At first people think it is a hoax. Soon the invention makes Edison famous.

1895 Guglielmo Marconi makes a device that uses an antenna to send messages over several kilometers without the use of wires. Fourteen years later a live opera is broadcast on radio. By the 1920s, many fans listen to weekly shows on home radios.

1898 Building on Edison's phonograph idea, Valdemar Poulsen patents the first tape recorder. By the 1940s, it is used by record companies. Ten years later it makes its way into homes.

The first commercial motion picture is shown in the U.S. The motion picture camera evolved from many nineteenth century inventions concerning light and motion.

1924 John Baird uses an empty biscuit box, knitting needles, an old motor, and the lens from a bicycle light to make a picture on a screen. Working in his attic, he transmits the shadowy image of a neighborhood boy from his transmitter in one room to his screen in another. Baird has made an early version of the television set.

1957 The first artificial satellite is launched into space. More follow. Soon TV stations bounce signals off satellites to satellite dishes at homes.

1969 Students and staff from two major colleges in the U.S. connect computers for the Defense Department. The Internet develops out of this networking idea.

1971 Ted E. Hoff invents the microchip. Working in a garage a few years later, Steven Jobs and Stephen Wozniak use the tiny processing chip in the first home computers.

1979 The first cellular phone system is established in Tokyo. Large towers transmit radio signals so the phones need no wires.

1980 Fax machines and scanners become available. A century earlier, Alexander Bain in Scotland sent the first drawing over telephone lines.

1983 Compact disc (or CD) digital recording is introduced. Today almost all music is produced on CDs.

1989 Tim Berners-Lee creates the World Wide Web which enables computer users to access a wide variety of information. People who live far apart suddenly seem a part of the same community. News from around the world becomes available instantly. People study, shop, conduct business, and talk to friends without leaving their homes.

Questions about Communicating Through the Ages

Answer the following questions in complete sentences.

1. Many printing presses were in use before Gutenberg's machine. What made his press special?

_____

2. What material is found inside a pencil?

_____

3. What machine did Latham Sholes patent?

_____

4. What was unusual about how Bell happened to invent the telephone?

_____

_____

_____

5. Who invented the phonograph?

_____

6. Where was the first home computer built?

_____

7. How has Tim Berners-Lee's invention affected you?

_____

_____

_____

8. How can the expression "It's a small world" be applied to advances in communication?

_____

_____

_____

Vocabulary

Complete the sentences below by filling in each blank with one of the words from the Word Box.

Word Box				
communicate	Internet	graphite	electric current	antenna
telegraph	transmitted	phonograph	radio waves	

1. Although many people think it is lead, it is really _____ that you find inside a pencil.

2. We _____ with one another when we exchange information.

3. Samuel Morse learned that _____ _____ could be sent over a wire.

4. A _____ is also sometimes called a record player.

5. Cellular phones need no wires because towers transmit _____ _____ over a long distance without wires.

6. The _____ is a system that connects computers to one another.

7. Coded messages are sent over long distances with the use of the

   _____ .

8. An _____ sends and receives radio waves.

9. A message that has been sent from one place to another has been

   _____ .

Toys Time Line

Countless innovations have changed the way we communicate. New ideas have also changed the way we play. Complete the time line of toys at the bottom of the page by writing the name of the invention that corresponds with each date.

Invention	Inventor	Date
Balloon	Michael Faraday	1824
Barbie Doll	Ruth Handler	1959
First Bicycle with Pedals	Kirkpatrick Macmillan	1839
Comic Strip	Richard Felton Outcault	1895
Crayola Crayons	Edwin Binney and Harold Smith	1903
Frisbee	Walter Morrison	1947
Game Boy	Gunpei Yokoi	1989
Jigsaw Puzzle	John Spilsbury	1767
Modern Roller Skates	J. L. Plimpton	1863
Duncan Yo-Yo	Donald Duncan	1920

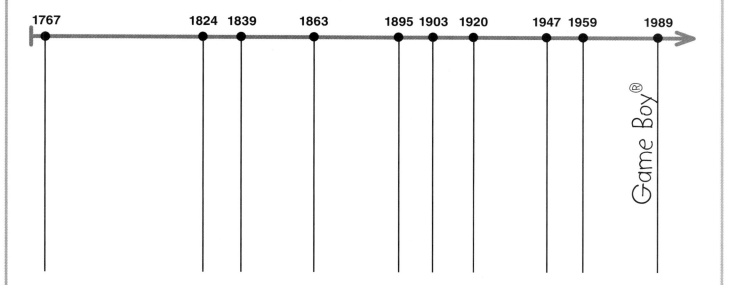

1767   1824 1839   1863   1895 1903 1920   1947 1959   1989

Game Boy®

Answer Key

Page 7

1. Answers will vary, but should reflect the following points:
   - The virus is transmitted from one person to another.
   - It enters the body through the nose (some students may infer that it can also enter through the mouth).
   - It attaches itself to a cell at the back of the throat.
2. The cell becomes infected and dies.
3. The body makes extra mucous to wash away the virus.
4. Coughing and sneezing are ways that the body tries to get rid of the virus.
5. Answers may vary, but should reflect the following points:
   - The purpose of antibodies is to fight the virus.
   - Once you have had a virus, the body knows how to make antibodies quickly that will fight off that particular virus.
6. You can wash your hands often and you can avoid sharing foods with people who have colds.

Page 8

A. 1. unhappy
2. easy-going
3. settles on
4. kill
5. responds
6. stick to

B. 1. cell
2. mucous
3. infected
4. receptor
5. antibody
6. particle

Page 9

Stories will vary, and may be fiction or nonfiction. Stories should describe the onset of a cold and have a clear beginning, middle, and end.

Page 12

1. Clams are hard to open because their shells are held closed by a powerful muscle.
2. Clams filter food from the water through a tube called a siphon.
3. The sea star wraps its arms around the clam and attaches itself with tiny suction cups. The arms pull on the two halves of the clam.

4. Barnacles have a special kind of cement that attaches them to rocks and other surfaces.
5. Hermit crabs move into larger shells as they outgrow their old ones.
6. Tide pools form when seawater remains trapped in rocks and crevices when the tide goes out.

Page 13

1. sea star
2. rockweed
3. clam
4. sea lettuce
5. hermit crab

Page 14

Page 17

1. Potential energy is the potential to do something.
2. Kinetic energy is the energy of motion.
3. a. The ball dents slightly. Its molecules stretch and bend, storing energy.
   b. The energy that caused the ball to fall is stored as potential energy.
   c. The floor also dents slightly, storing energy for a brief moment.
   Some responses may include what happens as the ball bounces upward as well:
   The ball and the floor return to their original shape. The stored energy becomes kinetic energy, as the ball bounces upward.
4. When the egg hits the floor, its molecules do not stretch and bend. The egg smashes.

Page 18

Students' sentences will vary, but should demonstrate understanding of word meaning.

Page 19

Although observations will differ, student conclusions should indicate that harder surfaces return more energy to the ball, causing it to bounce higher.

Page 22

1. The poem is mainly about two kinds of energy from the Sun: light and ultraviolet radiation.
2. Radiant energy is energy that travels out in all directions.
3. People can see light energy.
4. Photosynthesis is a process in which plants use light from the Sun to help them convert chemicals from the air and the ground into food for the plant.
5. Ultraviolet energy is a kind of radiant energy that comes from the Sun. People cannot see it. Ultraviolet waves are shorter than light waves.
6. Most of the ultraviolet waves from the Sun are blocked by Earth's upper atmosphere. Some ultraviolet radiation also gets lost out in space.
7. Ultraviolet radiation causes sunburn, wrinkles, and even cancer.

Page 23

Students' sentences will vary, but should contain homophones that are spelled correctly.

1. sun
2. we'll
3. see
4. to or two
5. right
6. here

Page 24

Venn diagrams will vary. Possible facts include:

Light
Helps plants grow
Can be seen by people
Longer waves
Lots of light reaches Earth's surface

Both
Come from the sun
Are forms of radiant energy

Ultraviolet Energy
Causes sunburn, wrinkles, cancer
Travels in waves
Cannot be seen by people
Waves are shorter
Very little reaches Earth's surface

Page 27
1. a pulling force that exists between all objects
2. exert greater gravity on each other
3. as the objects move farther apart
4. the force of gravity between Earth and the Moon
5. the Moon has less mass and thus less gravitational pull

Page 28
1. solar system
2. bulges
3. rotates
4. ebb, flow
5. dense
6. exert
7. orbit
8. Mass
9. affects
10. exists

Page 29
Paragraphs will vary, but should demonstrate understanding of the concept of gravity.

Page 32
1. at the same angle in which it hits an object
2. Your image is reversed twice, so it looks normal.
3. refraction
4. light rays bend when they move from water to air or air to water
5. It bends.
6. They both deal with light waves.

Page 33
1. refraction
2. reflection
3. Curved
4. regular
5. reversed
6. angle

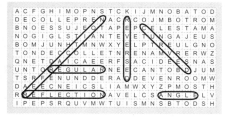

Page 34
Experiment 1—reflected
Experiment 2—refracted or bent
Experiment 3—higher, reflected, refracted or bent

Page 37
1. too complicated, too many parts, not practical
2. a wax cylinder, needle, vibrations, amplified

3. The sketch shows how parts will connect and helps you figure out the assembly.
4. A working model is the semi-final assembly, after the bugs and problems have been worked out. The prototype is the finished invention as it will look and function when it is actually manufactured.
5. The breadboard will show Jason if his invention works. It is made of whatever materials and parts are available. There would be no point in building a working model unless it actually does work.

Page 38
1. facial tissues
2. gelatin dessert
3. cellophane tape
4. copying by machine
5. adhesive bandage
6. ice pop
7. inline skates
8. cotton swabs
Gizmos Unlimited—The dictionary defines a gizmo as a gadget, but students' definitions will vary. Accept reasonable responses.

Page 39
Students' "inventions" will vary, but each must begin with an existing object and add 10 parts; all parts are to be labeled with an explanation of their function.

Page 42
1. Yeast makes bread dough rise.
2. *All Rise* tells what yeast is, how it was discovered, and how it is used today.
3. Yeast absorbs food from other substances.
4. Louis Pasteur baked the first loaf of bread.
5. Wet and dry yeasts are used in baking.

Page 43
1. absorb
2. substance
3. fortify
4. enzyme
5. fermentation
6. fungus
7. reproduce
8. carbon dioxide
9. chlorophyll
10. organism
11. semi-dormant
12. budding

Page 44
Works That Saved People's Lives
showed how to kill germs by heating food
told doctors to wash hands and instruments
created a treatment for rabies
discovered bacteria that cause human disease
built a rabies clinic
Works that Saved People's Businesses
helped wine makers make better wine
saved silk industry
saved farmers' animals with vaccines

Page 47
1. The first car ever built was powered by steam.
2. Solar power is a renewable energy source.
3. A solar-powered car would be most practical in places that get a lot of sun year round, and also places where the distances to travel are not great.
4. biomass fuel
5. Alike—need to be recharged frequently, take about three hours to recharge, run on battery

   Different—solar powered by sun, electric powered by power plant electricity
6. Solar and electric cars are not good for long-distance driving because they need to be recharged so often.
7. Students' opinions will vary, but should be supported.

Page 48
1. hybrid
2. alternative energy, renewable energy
3. solar power
4. electricity

Page 49
Take a Bus or Subway
Advantages—save energy, someone else drives, can accomplish things while traveling

Disadvantages—can be crowded, have to follow someone else's schedule, only makes stops at designated places

Ride a Bike

Advantages—save energy, get exercise, get to enjoy nature, doesn't pollute

Disadvantages—not safe to ride a bike on all roads; have to dress appropriately, so may need to change when you get to destination; impractical for long distances

Walk

Advantages—same as bike

Disadvantages—takes longer to get to destination, impractical for long distances, sidewalks not always available where you need to go

Page 52
1. ice and rock
2. reflect the sun's light
3. the solar wind blows dust and vapor away from the comet
4. oval-shaped orbits
5. takes less than 200 years to make its orbit

Page 53
A. Drawings and labels should show that the nucleus is the center of the comet, that the coma surrounds the nucleus, and that the tail is part of the comet that points away from the sun.
B. Students' new words and sentences should demonstrate understanding of the following base words:
   2. approach   4. tilt
   3. reflect   5. figure

Page 54
First time line:
1992—Comet Shoemaker-Levy was found to be in a very close orbit around Jupiter.
1997—Comet Hale-Bopp appeared in Earth's sky.
2061—Halley's comet is expected to return.
Bonus Question: Jan was born in 1992. We know this because she was 5 years old when she saw comet Hale-Bopp, which appeared in 1997.
Second time line:
Events shown on the time line should demonstrate understanding of how time lines are used.

Page 57
1. sedimentary rock
2. Sedimentary rock means "rock made from fire."
3. Metamorphic rock is made from the two other kinds of rock.
4. sunlight
5. It is used to make lime-flavored drinks.
6. made from fire

Page 58
1. T
2. F Slate is a metamorphic rock.
3. T
4. F Granite is an igneous rock.
5. T
6. T
7. F When tiny pieces of shell get pressed together, they form sedimentary rock.
8. T
9. F When limestone gets hot, it yields lime.
10. F Jewelry is made from quartz.

Page 59
1. basalt   4. limestone
2. pumice   5. conglomerate
3. sandstone   6. obsidian

Page 62
1. new moon
2. The Moon passes between the Sun and Earth.
3. an apple
4. during totality
5. The Moon appears to be the same size as the Sun because it is closer to Earth.

Page 63
A. 1. not hurried
   2. half a circle
   3. not ripe
   4. two coasts
   5. partial or half darkness
   6. two colored, or in two colors
   7. having two poles
   8. not in season
B. 1. bicolored
   2. unseasonable
   3. unhurried
   4. semidarkness
   5. semicircle
   6. bicoastal
   7. bipolar
   6. unripe

Page 64
Students' "news stories" will vary, but should correctly describe a solar eclipse.

Page 67
1. Peas and carrots are a mixture because both vegetables retain their individual properties, and the peas can be separated from the carrots with little difficulty.
2. Answers will vary. Sample answers: Cracker Jack, vegetable soup, sand and gravel, a can of mixed nuts, a tossed salad
3. water   sodium + hydrogen + chlorine + oxygen
   salt   hydrogen + oxygen
   bleach   sodium + chloride
4. Bleach and ammonia are both compounds. One of the elements in bleach is chlorine, safe when combined, but when Osgood added the bleach to the ammonia, he formed a new compound, and the chlorine in the bleach was released as a poisonous gas. Osgood was killed when he inhaled the chlorine gas.

Page 68
1. mixture   5. ammonia
2. hydrogen   6. properties
3. salt   7. oxygen
4. chlorine   8. compound

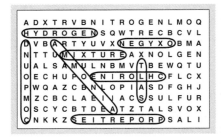

Page 72
1. cumulonimbus
2. spinning winds that form inside supercells
3. in Kansas
4. the red geraniums it had sucked up
5. Mixing air masses create the right conditions.

Page 73
A. 1. a color
   2. the place where air masses meet
   3. a notice or bulletin
   4. pulling upward
B. The tornado rushed forward, swallowing the barn.
   inhaling everything
   the funnel blushed
   The tornado pulled at the storm cellar doors, wrenching them free and carrying them away
   it bellowed like a T-Rex

Page 74
2
1
5
3
4

Page 77
1. numerous purple wildflowers
2. Locked sections of the fault do not move for a hundred or more years at a time.
3. Plates creep at the rate of one mile per year.
4. locked sections of a fault moving suddenly and sending out vibrations
5. thousands of tiny earthquakes
6. The Southern California earthquake occurred in an unpopulated area.

Page 78
Across
 2. earthquake
 3. crust
 5. disaster
 6. fault
 7. Pacific plate
 8. tectonic plates
 9. San Andreas
10. geologist

Down
 1. vegetation
 4. San Francisco

Page 79
1. T   4. F
2. T   5. F
3. T   6. T

Page 82
1. thrust
2. 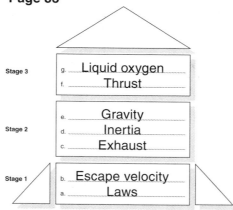 exhaust → up
3. It supplies oxygen so the fuel can burn.

4. His escape velocity will be less than 7 miles per second.

Page 83

Stage 3 — g. Liquid oxygen
          f. Thrust

Stage 2 — e. Gravity
          d. Inertia
          c. Exhaust

Stage 1 — b. Escape velocity
          a. Laws

Page 84
Answers will vary. Suggested answers are:
Letter 1—Yes, one of Newton's Laws of Motion says, "An object at rest tends to remain at rest." The weight of your rocket requires a lot of force to overcome the gravity that holds it to the ground.
Letter 2—The amount of fuel you used did not provide enough thrust to get your rocket to escape velocity (7 miles per second).
Letter 3—Just remember what Newton said, "An object in motion remains in motion unless something acts upon it to slow it down or speed it up."
So don't worry, you're not going to slow down because there is nothing in space to slow you down.

Page 87
1. strung wires and antennae between tall poles
2. The signals showed up as marks on chart paper.
3. She waited to see if it would happen again.
4. a new kind of star called a pulsar
5. are good timekeepers

Page 88
1. astronomy   6. dense
2. solar system   7. antenna, antennae
3. pulsar   8. collapse
4. pulses   9. regular
5. radio telescopes

Page 89
1. Answers will vary, but should indicate areas of student interest.
2. Stories may be fiction or nonfiction. Stories should be on topic and contain a minimum of errors in spelling, punctuation, and capitalization.

Page 92
1. Elements are the basic building blocks of all things.
2. The periodic table is a chart that arranges elements by atomic number.
3. Dmitry Mendeleev published the first version of the periodic table.
4. Each square of the periodic table includes the name of the element, the symbol of the element, and the element's atomic number and atomic mass.
5. A group or family is made up of vertical columns of closely related elements.
6. Chemists use the periodic table to distinguish metals from nonmetals, to check the mass and symbol of elements, and to know which elements share traits.

Page 93
1. vertical
2. electrons
3. atomic weight
4. proton
5. periods
6. periodic table
7. symbol
8. atomic number
Letters with **x**: l, t, e, n, s, e, m, e
Unscrambled: elements

Page 94
1. 21
2. Ne
3. Any three of the following: Al, Ge, Sb, Po, B, Si, As, Te, At
4. Au
5. Rn
6. F, Cl, Br, I, At

Page 97
1. He was an orphan without a high school education.
2. He helped the British mass-produce a radar device that helped them locate enemy ships in the dark.
3. They melted a candy bar in his shirt pocket.
4. Radaranges were first used in factories, restaurants, ships, and railroad cars.
5. They were too big and expensive and needed special plumbing.
6. They thought microwaves could harm them or their food.
7. They move molecules around quickly to create heat.

Page 98
A. 1. c–microwave
   2. d–radar
   3. e–patented
   4. a–magnetron
   5. b–orphaned
   6. f–molecules
B. Sentences will vary. Accept all reasonable responses.

Page 99
1. can opener
2. mixer
3. dishwasher
4. refrigerator
5. toaster
6. blender
7. coffee maker

Page 102
1. balsa wood
2. He discovered it through trial and error.
3. on the wing of the plane
4. It increased lift to keep the plane in the air.
5. an inventor

Page 103
A. These words appear in the story and should be circled:
   nose dives   brainstorm
   livestock   afternoon
   Students should have written these additional compound words:
   snowstorm scatterbrain stockpile
   nosebleed

B. 1. earmuffs
   2. airbag
   3. water-skiing

Page 104
Students' inventions will vary.

Page 107
1. become warmer
2. trap heat near Earth's surface
3. Air conditioners, cars, and factories produce greenhouse gases.
4. Plants would be smaller.
5. all of the above

Page 108
1. migrate   8. climate
2. technology   9. solar energy
3. carbon dioxide   10. atmosphere
4. global warming   11. fossil fuels
5. cycle   12. pollute
6. dozen   13. mined
7. constant   14. research

Page 109
1. Chapter 2
2. pages 38 through 44
3. Chapter 1
4. page 79
5. The Table of Contents tells you how the book is organized, and gives the beginning page number for each section or chapter.

Page 112
1. a cluster of cells that can detect light and dark
2. It helps the animal to see if something is sneaking up on it.
3. judge distance and see motion
4. the ability to see well in the dark
5. Vision will not be blocked by a large shell.

Page 113
A. 1. simple
   2. sense
   3. movement
   4. determine
   5. form
B. 1. complex
   2. prey
   3. duller
   4. ordinary
C. These words should be circled:
   dependent   visible   sensitive

Page 114
Students' sketches and descriptions will vary, but should show an understanding of adaptations needed for each of the two habitats described.

Page 117
1. Human actions have caused all animal extinctions.
2. Vacationing in the African wild is an adventure.
3. Vacationing in the African wild exploits animals and Native African cultures.
4. spend enough money
5. all of the above

Page 118
1. climate   6. evolution
2. extinction   7. trophy
3. natural disasters 8. tourism
4. natural resources 9. marvel
5. poacher   10. predator

Page 119
1. China   4. blue whale
2. 2,500   5. tiger
3. India   6. United States

Page 122
1. The air is clear and few people live in the area, so there is little "light pollution."
2. A reflector telescope uses mirrors to collect light from objects in space, making the objects seem brighter and closer than they are.
3. The Harlan-Smith telescope helped scientists prepare for the Viking mission to Mars.
   Scientists learned about the atmospheres of distant planets, and how far away the planets were. They also used the telescope to decide what they wanted to learn more about, sort of like we might use a guide book to learn about a place before we vacation there.
4. The McDonald Lunar Ranging Station shoots laser beams at reflectors on the Moon. The beams bounce off the reflectors and return to Earth. Scientists record how long this takes in order to measure movement of the Moon very precisely.